ACT Math

by

SGK Teaches

To my lovely & wonderful wife
My inspiration and support
My best friend
Smitha Siravuri

About the Author

Srinivasa Sastri Siravuri offers very unique and innovative ways of teaching mathematics. Siravuri's academic qualifications include a Ph.D. in Engineering. Siravuri has many years of experience teaching mathematics. Throughout this book, Siravuri offers many time saving tips, short cuts and tricks to solving the math problems.

Student's feedback

Mr. Siravuri is a brilliant, dedicated, and very encouraging teacher. He demonstrates the most efficient ways to approach a problem. As time management was an issue for me, this aspect of the book was very helpful. ~Chitra

Mr. Siravuri teaches multiple techniques and shortcuts for effective time management and uses lots of practice problems to reinforce the concepts. In the math sections, there are many ways to quickly eliminate several options. His passion is evident while he teaches. ~Swathi

From the Author

I benefited from my Dad, who is a great math teacher. This is my humble attempt to pass on my learnings to the students. My efforts need your encouragement and support. Please feel free to write to sgk@sgkteaches.com.

Website: www.sgkteaches.com

E-mail: sgk@sgkteaches.com

ISBN-13: 978-1512376050

ISBN-10: 1512376051

Library of Congress Control Number: 2015912519

Contents

Introduction

Standardized tests are not easy to crack. Knowing the concepts will not ensure success in the standardized tests. Applying the concepts is even more important than knowing the concepts. Sometimes a problem involves applying multiple concepts as illustrated in this book.

- Throughout the book, look for *"SGK's Short Cut"*. The solution outlines the best strategy for a given problem.
- Always read the answer choices first. You may be able to pick the best answer without actually solving the problem. It will come with practice. Practice helps you with time management.
- Do not skip a chapter or a problem even if you are familiar with the concept. Review the chapter so that you learn new and better ways of solving problems.
- Some problems can be solved *without using a calculator*. Recent changes to the standardized tests might require you to solve problems without using a calculator.

Review the following problems to understand what this book is about.

<u>Example 1</u>: If $n^2 + 12 = 93$, what is $n^2 - 1$?

    A. 100   B. 94   C. 81   D. 80   E. 10

<u>One method that is not recommended is</u>

If $n^2 + 12 = 93$, $n^2 = 93 - 12 = 81$. Taking a square root yields, $n = \sqrt{81} = 9$.

Substituting n=9, $n^2 - 1 = 9^2 - 1 = 80$.

<u>*SGK's Short Cut:*</u>

$n^2 + 12$ is greater than $n^2 - 1$. Hence, the answer should be less than 93. Eliminate choices A and B. Choice E is too small.

Note that both the terms $n^2 + 12$ and $n^2 - 1$ contain n^2 as a <u>common term.</u>

The difference between $n^2 + 12$ and $n^2 - 1 = 12 - (-1) = 13$.

Hence, $n^2 - 1 = 93 - 13 = 80$. In other words, you do not need to solve for n. You also do not need a calculator to solve this problem.

Once you recognize $n^2 = 81$, calculate $n^2 - 1 = 81 - 1 = 80$.

<u>Example 2</u>: If $h(x) = -9x^2$, what is $h(7)$?

    A. 567   B. 639   C. -441   D. 63   E. -561 h(x)=-9x^2; substituting $x = 7$ yields, $h(7) = -9 \times 7 \times 7 = -441$.

SGK's Short Cut:

The concept being tested here is that x^2 is always positive for all real values of x. Hence, f(x)=-9x^2 is always negative. Hence, you can eliminate choices A, B and D. You are also looking for a number that is divisible by **9** because f(x)=**-9**x^2. In looking at choices C and E, -441 is divisible by 9 and -561 is not. A number is divisible by 9 if the sum of the digits is divisible by 9. When we take -441, note that 4+4+1=9 which is divisible by 9. When we take -561, note that 5+6+1=12 which is not divisible by 9. Hence, we can eliminate choice E although it is a negative number. Hence, choice C is the only possible answer. You do not need a calculator to solve this problem.

<u>Example 3</u>: If $\frac{y}{27} = 3^t$, what is the value of y when $t = 4$?

    A. 270   B. 1550   C. 2187   D.2200   E. 2790

One method that is not recommended is

Rewriting the equation $\frac{y}{27} = 3^t$ yields $y = 27 \times 3^t$. Substituting t=4 in the equation yields, y=2187.

SGK's Short Cut:

The concept being tested here is <u>multiple of 10</u>. 10=2×5. Hence, any number that is a multiple of 10 (numbers ending in a zero) should be a <u>multiple of 5</u>. However, note that in the given equation, $\frac{y}{27} = 3^t$; $y = 27 \times 3^t$. In the equation, y does not involve <u>a multiplication with 5</u>. Hence, the answer <u>cannot end in a zero</u>. Eliminate all choices <u>except C.</u>

<u>Example 4</u>: What is the slope of the line 13x+17y=100?

One method that is not recommended is

Rewrite the equation as 17y=100-13x; dividing by 17 yields, $y = \frac{-13}{17}x + \frac{100}{17}$.

Comparing the equation to y=mx+c yields, the slope of the equation = $\frac{-13}{17}$.

SGK's Short Cut:

The slope of the equation ax+by+c=0 is $\frac{-a}{b}$. Hence, the slope of the given equation 13x+17y=100 is $\frac{-13}{17}$.

ax+by+c=0 is called the general form of a straight line. To find out the slope, calculate $\frac{-x\ \text{coefficient}}{y\ \text{coefficient}}$.

See the table below.

Equation in the general form	X-Coefficient	Y-Coefficient	The slope of the line
2x+3y+4=0	2	3	$\dfrac{-2}{3}$
-3x+5y+9=0	-3	5	$\dfrac{3}{5}$
8x-2y+12=0	8	-2	4
4x-7y+13=0	4	-7	$\dfrac{4}{7}$

Example 5: Simplify the expression $\dfrac{12+\frac{1}{7}}{2+\frac{1}{5}}$.

    A. $3\frac{1}{5}$ B. $4\frac{2}{35}$ C. $7\frac{3}{7}$ D. $5\frac{40}{77}$ E. $10\frac{5}{7}$

One method that is not recommended is

$12 + \dfrac{1}{7} = \dfrac{12\times7+1}{7} = \dfrac{85}{7}$.

$2+\dfrac{1}{5} = \dfrac{2\times5+1}{5} = \dfrac{11}{5}$.

$\dfrac{12+\frac{1}{7}}{2+\frac{1}{5}} = \dfrac{\frac{85}{7}}{\frac{11}{5}} = \dfrac{85}{7} \times \dfrac{5}{11} = \dfrac{425}{77} = 5\frac{40}{77}$.

SGK's Short Cut:

By noting that $\dfrac{12}{2} = 6$, we are looking for an answer close to 6 but less than 6 because $\dfrac{1}{7}$ in the numerator is less than $\dfrac{1}{5}$ in the denominator. Hence, choices C and E can be eliminated. Choices A and B are too low and hence, can be eliminated as well. <u>Choice D is the only answer that is <u>close to 6 but less than 6.</u>

Example 6: What is the remainder when $f(x) = 5x^3 + 6x^2 - 8x + 1$ is divided by $x - 1$?

    A. 5   B. 6   C. 4   D. 7   E. 8

One method that is not recommended is

Use the long division or synthetic division to find the remainder. Both long division and synthetic division take a long time to solve this problem.

When a polynomial function f(x) is divided by x-a, the remainder is f(a). Hence, substitute x=1 in the equation or find out f(1). f(1) = 5+6-8+1=4. Hence, 4 is the remainder.

If you are impressed with these examples, continue to the rest of the book to learn many more techniques.

1. Properties of Numbers

Knowing the number types is very important to answer certain math questions. Some questions specify the type of number involved in the problem.

Natural Numbers

Natural numbers are also called counting numbers. Natural numbers start with 1. e.g.: 1, 2, 3, 4.

Whole Numbers

Whole numbers start with 0. The set of whole numbers includes zero and natural numbers. e.g.: 0, 1, 2, 3, 4. Note that both natural numbers and whole numbers do not include negative numbers.

Integers

Integers are <u>both positive and negative</u>. 0 is included in the set of integers. e.g.: …,-3,-2,-1, 0, 1, 2, 3.

- Positive integers are 1, 2, 3, …
- Negative integers are -1, -2, -3, …
- *Zero is neither positive nor negative.*

Odd Integers

Integers not divisible by 2 are called odd integers or odd numbers. e.g.: …, -3, -1, 1, 3, …

Even Integers

Integers divisible by 2 are called even integers or even numbers. e.g.: …, -4, -2, 0, 2, 4, …

As you notice 0 is divisible by 2 and hence, zero is an <u>even number</u>.

Prime Numbers

A prime number is a number that is not divisible by any number other than 1 and itself. e.g.: 2, 3, 5, 7, 11. Note that

- 1 is <u>not</u> a prime number.
- 2 is a prime number; 2 is the smallest prime number. 2 is the only *even prime number*. The rest of the prime numbers are odd. For this reason, the sum of two prime numbers can be even or odd.
    - If the sum of two prime numbers is odd, one of them must be 2. Similarly, if the product of two prime numbers is even, one of them must be 2.
    - If the sum of two prime numbers is even, both the numbers are odd and hence, 2 is not one of them.

Prime Factorization

A *factor* is a number that divides the number evenly (with zero as the remainder). Prime factorization expresses a number as a product of prime numbers.

<u>Example 1</u>: Find the prime factorization for numbers 12 and 144.

$12 = 2 \times 2 \times 3 = 2^2 \times 3$.

$144 = 2 \times 2 \times 2 \times 2 \times 3 \times 3 = 2^4 \times 3^2$.

Prime factorization is unique in that there is <u>one and only one</u> way to express a number using prime factorization. Prime factorization is a powerful concept and it can be used to solve a variety of problems as demonstrated below.

Applications of Prime Factorization

<u>Example 2</u>: Simplify the ratio $\frac{144}{972}$.

$972 = 2 \times 2 \times 3 \times 3 \times 3 \times 3 \times 3 = 2^2 \times 3^5$.

$$\frac{144}{972} = \frac{2^4 \times 3^2}{2^2 \times 3^5} = \frac{2^{4-2}}{3^{5-2}} = \frac{2^2}{3^3} = \frac{4}{27}.$$

In 3^5, 3 is called the base and 5 is called the exponent. Exponents are discussed in more detail in Chapter 7.

<u>Example 3</u>: $\frac{5}{12} - \frac{3}{45} = ?$

The problem requires us to find the common denominator for 12 and 45.

Here is how we use the prime factorization to find the common denominator.

$12 = 2 \times 2 \times 3 = 2^2 \times 3$.

$45 = 3 \times 3 \times 5 = 3^2 \times 5$.

Note the prime numbers involved in the prime factorization are 2, 3 and 5. To find the common denominator, we take each of the prime numbers with the highest power, i.e., $2^2, 3^2$ and 5. Multiplying these will give us $4 \times 9 \times 5 = 180$ as the common denominator.

$\frac{5}{12}$ and $\frac{75}{180}$ are equivalent fractions because 5×180=12×75.

Similarly, $\frac{3}{45}$ and $\frac{12}{180}$ are equivalent fractions because 3×180=45×12.

$$\frac{5}{12} - \frac{3}{45} = \frac{75}{180} - \frac{12}{180} = \frac{75-12}{180} = \frac{63}{180} = \frac{9 \times 7}{9 \times 20} = \frac{7}{20}.$$

Place Value and Decimal System

"Deci" means 10. Number system is based on 10 digits, 0 thru 9; hence, it is called the decimal system.

The number 23,758 is used to illustrate the decimal system.

Ten Thousands	Thousands	Hundreds	Tens	Units
10,000	1,000	100	10	1
10^4	10^3	10^2	10^1	10^0
2	3	7	5	8
2×10000	3×1000	7×100	5×10	8×1
20,000	3,000	700	50	8

20,000+3,000+700+50+8=23,758.

Here is another example using the number 0.3584.

Tenths	Hundredths	Thousandths	Ten Thousandths
$\dfrac{1}{10}$	$\dfrac{1}{100}$	$\dfrac{1}{1000}$	$\dfrac{1}{10000}$
10^{-1}	10^{-2}	10^{-3}	10^{-4}
3	5	8	4
$\dfrac{3}{10}$	$\dfrac{5}{100}$	$\dfrac{8}{1000}$	$\dfrac{4}{10000}$
0.3	0.05	0.008	0.0004

0.3+0.05+0.008+0.0004=0.3584.

Tens digit is not same as tenths digit. Hundreds digit is not same as the hundredths digit.

Application of the Decimal System

Example 4: Find x-y if x and y equal hundreds digit and tenths digit of 23845.678 respectively?

Note that the hundreds digit is 8 and tenths digit is 6 (not to be confused by tens digit which is equal to 4). The difference = x-y=8-6 = 2.

Rules of Addition

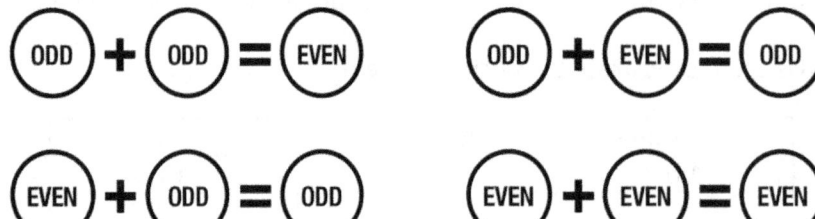

Rules of Multiplication

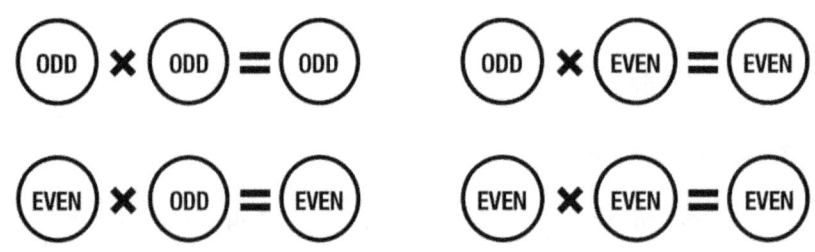

Special Cases

If X is odd, is X^2 odd or even? Is X^n odd or even? Based on the above rules, if X is odd, X^2 and X^n are odd.

$$\text{ODD} \times \text{ODD} = \text{ODD}$$

If X is even, Is X^2 odd or even? Is X^n odd or even? Based on the above rules, if X is even, X^2 and X^n are even.

$$\text{EVEN} \times \text{EVEN} = \text{EVEN}$$

Divisibility Rules

"Divisible" means the remainder is zero.

Divisibility Rule for 2

- Any even number is divisible by 2.
  - i.e., any number ending in 0, 2, 4, 6, or 8. In other words, the unit's digit is 0, 2, 4, 6, 8.

- When a number is divided by "2", there are only two possibilities for remainder, "0" or "1".
  - When an even number is divided by 2, the remainder is 0.
  - When an odd number is divided by 2, the remainder is 1.

Divisibility Rule for 3

- If the sum of all digits is divisible by 3, then the number is divisible by 3.
  - e.g.: 1242. The sum of digits = 1 + 2 + 4 + 2 = 9. "9" is divisible by 3. So, 1242 is divisible by 3.
  - e.g.: 321677. The sum of digits = 3+2+1+6+7+7=26 = 2+ 6 = 8. "8" is not divisible by 3. So, 321678 is not divisible by 3.
- When an integer is divided by "3", there are three possibilities for remainder, 0 or 1 or 2.

Divisibility Rule for 4

- If the last two digits are divisible by 4, then the number is divisible by 4.
  - e.g.: 1244. 44 is divisible by 4, hence 1244 is divisible by 4.
  - e.g.: 321678. 78 is not divisible by 4, hence 321678 is not divisible by 4.
- 100 is divisible by 4, so for every 100, the same rule applies. Hence, inspecting last two digit is enough. Note that 100, 200, 300, 400.. are all divisible by 4.
- When an integer is divided by "4", there are four possibilities for remainder, 0 or 1 or 2 or 3.

Divisibility Rule for 5

- If the last digit is 0 or 5, then the number is divisible by 5.
  - e.g.: 124405 ends in 5. Hence, 124405 is divisible by 5.
  - e.g.: 321673 ends in 3. Hence, 321673 is not divisible by 5.
- When an integer is divided by "5", there are four possibilities for remainder, 0 or 1 or 2 or 3 or 4.

Divisibility Rule for 6

- If a number is divisible by both 2 and 3, then it is divisible by 6.
  - e.g.: 1244112 is divisible by both 2 and 3. Hence, 1244112 is divisible by 6.
  - e.g.: 3216783 is divisible by 3 but not by 2. Hence, 3216783 is not divisible by 6.
- Note all numbers divisible by 6 are even numbers.

Divisibility Rule for 7

- Multiply last digit by 2 and subtract from the rest of the number; repeat the process if needed.
  - 91; 9-(1x2)=9-2=7; hence 91 is divisible by 7.
  - 721; 72-(1x2)=72-2=70; 70 is divisible by 7; hence 721 is divisible by 7.
- Alternately, multiple each digit by the following:

Explanation	Hundred Thousands with	Ten Thousands with	Thousands with	Hundreds with	Tens with	Units with
Multiply by	5	4	6	2	3	1
Given Number is			1	7	5	0
			6×1	2×7	3×5	1×0
Add these numbers			6	14	15	0

- For example, consider the number 1750. 6+14+15+0=35 is divisible by 7. Hence, 1750 is divisible by 7.

Divisibility Rule for 8

- If the last three digits are divisible by 8, then the number is divisible by 8.
  - e.g.: 1248. 248 is divisible by 8. Hence 1248 is divisible by 8.
  - e.g.: 321678. 678 is not divisible by 8. Hence, 321678 is not divisible by 8.
- 1000 is divisible by 8, so for every 1000, the same rule applies. Hence, inspecting last three digits is enough. Note that 1000, 2000, 3000... are all divisible by 8.

Divisibility Rule for 9

- If the sum of all digits is divisible by 9, then the number is divisible by 9.
  - e.g.: 1242. The sum of digits = 1 + 2 + 4 + 2 = 9. "9" is divisible by 9. So, 1242 is divisible by 9.
  - e.g.: 321677. The sum of digits = 26 = 2+ 6 = 8. "8" is not divisible by 9. So, 321677 is not divisible by 9.
- Note that the sum of digits also equals the remainder when divided by 9.
  - e.g.: When 111 is divided by 9, the remainder is 1+ 1+ 1 = 3.

Divisibility Rule for 10

- If the last digit is 0, then the number is divisible by 10.
  - e.g.: 124400 is divisible by 10.
  - e.g.: 321671 is not divisible by 10. The last digit indicates the remainder.

Divisibility Rule for 11

- Add the alternate digits. If the difference between the two sums is equal to 0 or a multiple of 11, then the number is divisible by 11.
  - e.g.: 1243. The sum of alternate digits =2+3=5 =1+4. Hence, 1243 is divisible by 11.
  - e.g.: 321677. The sum of alternate digits =3+1+7=11 and 2+6+7=15. Hence, 321677 is not divisible by 11.

Problems Involving Consecutive Integers

Example 5: A counter is issuing integer numbers consecutively. The sum of the last three numbers issued by the counter adds up to 252. What is the last number issued by the counter?

Consecutive integers are 1,2,3,4,5,.... A typical approach is to assume that the first number is "X". Next number is "X+1", the last number is "X+2".

The sum of the 3 three numbers = X + (X+1) + (X+2) = 3X+3.

Form the equation as 3X + 3 = 252; solving for X, $X = \frac{252-3}{3} = 83$.

The largest integer is X+2 = 83+2=85.

However, there is an easier way to solve this problem. Assume the numbers to be X-1, X and X+1. The sum of the three numbers = X-1+X+X+1=3X. Note that the middle number can be found by dividing the sum of the integers with 3.

So, $X = \frac{252}{3} = 84$. The numbers are 83, 84 and 85.

SGK's Short Cut:

Use the following table for problems involving consecutive integers.

Problem describes	Assume Numbers as	Middle Number Is
3 consecutive integers	X-1, X and X+1;	X=Sum of the integers divided by 3.
5 consecutive integers	X-2, X-1, X, X+1, X+2	X=Sum of the integers divided by 5.
7 consecutive integers	X-3, X-2,X-1, X, X+1, X+2, X+3	X=Sum of integers divided by 7.
3 consecutive odd integers	X-2, X and X+2	X=Sum of the integers divided by 3.
5 consecutive odd integers	X-4,X-2,X,X+2,X+4	X=Sum of integers divided by 5.
3 consecutive even integers	X-2,X and X+2	X=Sum of the integers divided by 3.
5 consecutive even integers	X-4,X-2,X,X+2,X+4	X=Sum of integers by divided by 5.
4 consecutive odd integers	X-3,X-1, X+1,X+3	$X=\frac{Sum\ of\ Integers}{4}$; X is still the middle number.
4 consecutive even integers	X-3,X-1, X+1,X+3	$X=\frac{Sum\ of\ integers}{4}$; X is still the middle number.

Practice Problems

1. The smallest prime number is
   A. 3   B. 0   C. 1   D. 2   E. 5

2. My son gave me a list of prime factors of 42. The sum of these numbers equals what number?
   A. 42   B. 10   C. 12   D. 15   E. 13

3. If "n" is an integer, which of the following is <u>not</u> necessarily even?
   A. 2n   B. 4n+58   C. $6n^2 + 6n + 10$ D. n+34   E. 6n+34

4. Find x+y if x is the units digit and y is the tenths digits of 234.78 respectively?
   A. 7   B. 5   C. 11   D. 12   E. 15

5. X, Y and Z represent three consecutive integers. X+Y+Z is always divisible by the following number
   A. 5   B. 4   C. 3   D. 2   E. 6

6. A service counter at a resort issues tokens that are consecutive integers. The sum of the last three tokens issued is equal to 117. What is the last token issued?
   A. 20   B. 30   C. 38   D. 40   E. 39

7. Three siblings are each 2 years apart. The sum of their ages equals 153 years. How old is the youngest?
   A. 50   B. 51   C. 49   D. 53   E. 55

8. What is X if X represents the number of factors of 144, including 1 and 144?
   A. 10   B. 13   C. 15   D. 20   E. 8

9. What is Y if Y represents the number of factors of 98 <u>not</u> including 1 and 98?
   A. 3   B. 4   C. 5   D. 6   E. 7

10. What is Z if Z represents the number of factors of 243, including 1 and 243?
    A. 5   B. 6   C. 4   D. 3   E. 7

11. If X is a positive number and X < 1, X^4 is
    A. Greater than 1   B. Less than 1   C. 1   D. Odd   E. None of the above

12. Chocolates are 9 cents each. If Nathan pays $12, what is the change he gets back?
    A. 11 cents   B. 4 cents   C. 3 cents   D. 5 cents   E. no change

13. What is the remainder when 187634 is divided by 5?
    A. 2   B. 3   C. 4   D. 5   E. 6

14. What number when added to 867 will make it divisible by 9?
    A. 3   B. 6   C. 0   D. 5   E. 4

15. A family of 7 is assigned adjacent seats at a game. The numbers on the seats add up to 98. What is the smallest seat number?
    A. 11   B. 14   C. 17   D. 20   E. 25

16. A machine spits out consecutive odd integers. The last 4 numbers issued by the machine add up to 80. What is the last number issued by the machine?

    A. 21   B. 20   C. 17   D. 19   E. 23

17. A token system issues consecutive even integers. The sum of the last four tokens issued is 100. What is the last token issued?

    A. 22   B. 24   C. 26   D. 28   E. 30

Solutions to Practice Problems

1. **Best answer is D.** 0 and 1 are not prime numbers. 2 is the smallest prime number.

2. **Best answer is C.** $42 = 2 \times 3 \times 7$. The sum of the prime factors = 2 + 3 + 7 = 12.

3. **Best answer is D.** If "n" is an integer, 2n is even. 4n and 6n are even as well. Hence, choices A, B, C and E are even. Choice D is even or odd depending on whether n is even or odd.

4. **Best answer is C.** The unit's digit is 4 and the tenths digit is 7. The sum = 4 + 7 = 11.

5. **Best answer is C.** Let the integers be X-1, X and X+1. The sum is equal to 3X. Hence, sum of three consecutive integers is always divisible by 3. Most students make a mistake by picking 6. Note that the sum of 2, 3 and 4 is divisible by 3. The sum of 3, 4 and 5 is divisible by 6. The sum of 4, 5 and 6 is divisible by 3. So, the sum of three consecutive integers is divisible by 3; but <u>not</u> necessarily divisible by 6.

6. **Best answer is D.** The average of the three numbers $= \frac{117}{3} = 39$. 39 is the middle number. The numbers are 38, 39 and 40. The last token issued is 40.

7. **Best answer is C.** The average of the three numbers is $\frac{153}{3} = 51$. 51 is the middle number. The numbers are 49, 51 and 53. The youngest is 49 years old.

8. **Best answer is C.** $144 = 2 \times 2 \times 2 \times 2 \times 3 \times 3 = 2^4 \times 3^2$. Note that 4 and 2 are the exponents of 2 and 3 respectively. To find the total number of factors including 1 and 144, multiply the exponents by adding one to each of the exponents = (4+1) × (2+1) = 5 × 3 = 15.

9. **Best answer is B.** $98 = 2 \times 7 \times 7 = 2^1 \times 7^2$. To find the total number of factors including 1 and 98, multiply the exponents by adding one to each of the exponents = (1+1) × (2+1) = 2 × 3 = 6. Because we are excluding 1 and 98, the number of factors = 6-2=4.

10. **Best answer is B.** $243 = 3 \times 3 \times 3 \times 3 \times 3 = 3^5$. To find the total number of factors including 1 and 243, multiply the exponents by adding one to each of the exponents = (5+1) = 6.

11. **Best answer is B.** Note that X is between 0 and 1. Hence, any powers of X will be less than 1.

12. **Best answer is C.** The remainder of $\frac{1200}{9}$ is 3. *SGK's Short Cut:* The remainder when dividing by 9 is equal to the sum of the digits = (1+2+0+0)=3. So, 3 is the remainder.

13. **Best answer is C.** Remember the remainder is the unit's digit = 4.

14. **Best answer is B.** Add the digits = 8+7+6=21; add digits again 2+1=3. To be divisible by 9, the sum of the digits should be 9. Hence, add 6 (9-3=6) to make the number divisible by 9. Verify that 876+6=882 is divisible by 9.

15. **Best answer is A.** The average of the numbers $= \frac{98}{7}=14$. Hence, the middle number is 14. The numbers are 11, 12, 13, 14, 15, 16, and 17. The smallest seat number is 11.

16. **Best answer is E.** The average is $\frac{80}{4} = 20$. Hence, the numbers are 17, 19, 21 and 23.

17. **Best answer is D.** The average is $\frac{100}{4} = 25$. Hence, the numbers are 22, 24, 26 and 28.

2. Fractions

Fractions, ratios and percentages are similar in nature. We explore each of them in the next few chapters. A fraction has two components: _numerator_ and _denominator_.

$$\text{Fraction} = \frac{\text{Numerator}}{\text{Denominator}}.$$

Note that <u>the _denominator cannot be zero because division by zero is not defined_</u>.

For a fraction,

- If the absolute value of the numerator < the absolute value of the denominator; then the fraction is less than 1 and the fraction is called a proper fraction. Examples: $\frac{1}{2}, \frac{3}{4}, \frac{2}{5}$.

- If the absolute value of the numerator >= the absolute value of the denominator, then the fraction is greater than (or equal to) 1 and the fraction is called an improper fraction. Examples: $\frac{4}{3}, \frac{7}{3}, \frac{9}{5}$.

Equivalent Fractions

Two fractions are said to be equal _if and only if_ when you cross-multiply, they yield the same number. $\frac{1}{2}$ and $\frac{4}{8}$ are equivalent because 1×8=2×4=8. On the other hand, $\frac{1}{2}$ and $\frac{2}{5}$ are not equivalent because 1×5=5 and 2×2=4. This interesting concept can be applied to solve problems involving proportions or ratios.

Adding and Subtracting Fractions

Fractions can be added or subtracted easily if the denominator is the same. Therefore, adding and subtracting fractions involves finding a common denominator. It will be easier if you find the Least Common Denominator (L.C.D). Prime factorization can be used to find the Least Common Denominator.

<u>Example 1</u>: $\frac{1}{4} + \frac{3}{10} = ?$

First, find out the Least Common Denominator for 4 and 10.

Prime factorization of 4: 4 = 2×2 = 2^2

Prime factorization of 10: 10 = 2×5 = $2^1 \times 5^1$

Now, select only those that have the highest exponent for each of the prime factors. L.C.D. of 4 and 10 = $2^2 \times 5^1 = 2 \times 2 \times 5 = 20$.

$$\frac{1}{4} = \frac{5}{20}; \frac{3}{10} = \frac{6}{20}; \frac{1}{4} + \frac{3}{10} = \frac{5}{20} + \frac{6}{20} = \frac{5+6}{20} = \frac{11}{20}.$$

<u>Example 2:</u> $\dfrac{1}{x^3y^6} - \dfrac{3}{xy^2z^5} = ?$

To find out the Least Common Denominator, you can treat x, y and z as prime factors. Choose the terms that have the highest exponent. Hence, the Least Common Denominator = $x^3y^6z^5$.

$$\dfrac{1}{x^3y^6} - \dfrac{3}{xy^2z^5} = \dfrac{1}{x^3y^6}\cdot\dfrac{z^5}{z^5} - \dfrac{3}{xy^2z^5}\cdot\dfrac{x^2y^4}{x^2y^4} = \dfrac{z^5}{x^3y^6z^5} - \dfrac{3x^2y^4}{x^3y^6z^5} = \dfrac{z^5-3\,x^2y^4}{x^3y^6z^5}.$$

Multiplication of Fractions

When multiplying fractions, multiply numerators and multiply denominators to form a new fraction. To express the fraction in the simplest form, make sure common factors are canceled out.

$$\dfrac{1}{4} \times \dfrac{2}{3} = \dfrac{1\times 2}{4\times 3} = \dfrac{2}{12} = \dfrac{1}{6}.$$

$$\dfrac{x}{2}\,\dfrac{y}{3z} = \dfrac{xy}{2\times 3z} = \dfrac{xy}{6z}.$$

Division with Fractions

When dividing with a fraction, multiply the numerator with the reciprocal of the denominator.

$$\dfrac{\frac{1}{2}}{\frac{2}{3}} = \dfrac{1}{2}\times\dfrac{3}{2} = \dfrac{1\times 3}{2\times 2} = \dfrac{3}{4}.$$

Practice Problems

1. $\frac{2}{3} + \frac{2}{4} = ?$

   A. $\frac{4}{12}$　　　B. $\frac{4}{7}$　　　C. $\frac{7}{12}$　　　D. $\frac{7}{6}$　　　E. 30

2. Sam walked $\frac{2}{5}$ mile in the morning and $\frac{3}{7}$ mile in the evening. How many miles did he walk?

   A. $\frac{5}{12}$　　　B. $\frac{6}{35}$　　　C. $\frac{29}{35}$　　　D. $\frac{12}{5}$　　　E. $\frac{5}{35}$

3. What is the value of $\frac{-2}{3} \times \frac{3}{5}$?

   A. $\frac{6}{15}$　　　B. $\frac{-2}{5}$　　　C. $\frac{-5}{15}$　　　D. $\frac{2}{5}$　　　E. 30

4. What is $\dfrac{\frac{2}{7}}{\left(\frac{-3}{4}\right)}$?

   A. $\frac{-3}{14}$　　　B. $\frac{3}{14}$　　　C. $\frac{-8}{21}$　　　D. $\frac{-21}{8}$　　　E. $\frac{-14}{3}$

5. Adam won $\frac{3}{5}$ of the chess games in a tournament. If $\frac{1}{4}$ of the games were drawn, what fraction of the games did he lose?

   A. $\frac{7}{20}$　　　B. $\frac{3}{4}$　　　C. $\frac{2}{5}$　　　D. $\frac{4}{20}$　　　E. $\frac{3}{20}$

6. $\frac{1}{2} + \frac{1}{4} + \frac{1}{8} + \frac{1}{16} = ?$

   A. $\frac{15}{16}$　　　B. $\frac{1}{16}$　　　C. $\frac{7}{8}$　　　D. $\frac{3}{4}$　　　E. $\frac{31}{32}$

7. Simplify $\frac{1}{1+x} + \frac{1}{1-x}$

   A. $\frac{2}{1-x^2}$　　　B. $\frac{2x}{1-x^2}$　　　C. $\frac{-2x}{1-x^2}$　　　D. 0　　　E. 1

8. Simplify $\frac{1}{1-x} + \frac{1}{1-y}$

   A. $\frac{1}{1-x-y-xy}$　　B. $\frac{2-x-y}{1-x-y+xy}$　　C. $\frac{2-x-y}{1-x-y}$　　D. $\frac{1}{(1-x)(1-y)}$　　E. 1

9. Simplify $\frac{3}{x} + \frac{x+1}{x+5}$

   A. $\frac{x+4}{2x+5}$　　B. $\frac{x+4}{x(x+5)}$　　C. $\frac{3(x+1)}{x(x+5)}$　　D. $\frac{x^2+1}{x^2+5x}$　　E. $\frac{x^2+4x+15}{x(x+5)}$

10. Simplify $\frac{x^2-1}{x+1}$

    A. $x - 1$　　　B. $x + 1$　　　C. $\frac{1}{x+1}$　　　D. $\frac{1}{x-1}$　　　E. 1

11. Simplify $\frac{x^2-9}{x+3} \div \frac{x-5}{x^2-25}$

    A. $\frac{x-3}{x+5}$　　B. $\frac{x+5}{x-3}$　　C. $\frac{x+3}{x-5}$　　D. $x^2 + 2x - 15$　　E. 1

Solutions to Practice Problems

1. **Best answer is D.** _SGK's Short Cut:_ Choice E can be eliminated right away because it is too big. Note that choices A, B and C do not make sense either because they are proper fractions, and hence are less than 1. When you add $\frac{2}{3}$ and $\frac{1}{2}$, you expect an answer greater than 1. Hence, only choice D makes sense.

   Solving the problem traditional way, $\frac{2}{3} + \frac{2}{4} = \frac{2\times4+2\times3}{12} = \frac{8+6}{12} = \frac{14}{12} = \frac{7}{6}$.

2. **Best answer is C.** $\frac{2}{5} + \frac{3}{7} = \frac{2\times7+3\times5}{35} = \frac{14+15}{35} = \frac{29}{35}$.

3. **Best answer is B.** Eliminate choices A, D and E because we are expecting a negative number as an answer. If you have to guess, choose B or C. By canceling out 3, $\frac{-2}{3} \times \frac{3}{5} = \frac{-2}{5}$.

4. **Best answer is C.** $\frac{\frac{2}{7}}{\left(\frac{-3}{4}\right)} = \frac{2}{7} \times \frac{4}{(-3)} = \frac{2\times4}{7\times(-3)} = \frac{8}{-21} = \frac{-8}{21}$.

5. **Best answer is E.** Total games won, drawn and lost should add up to 1. Hence, fraction of games lost $= 1 - \left(\frac{3}{5} + \frac{1}{4}\right) = 1 - \left(\frac{3\times4+5\times1}{5\times4}\right) = 1 - \frac{12+5}{20} = 1 - \frac{17}{20} = \frac{20-17}{20} = \frac{3}{20}$.

6. **Best answer is A.** $\frac{1}{2} + \frac{1}{4} + \frac{1}{8} + \frac{1}{16} = \frac{8+4+2+1}{16} = \frac{15}{16}$.

7. **Best answer is A.** Here, I introduce a very powerful technique. It involves substituting values for x. Easy values to try are x=0 or x=1 or x=2. In this case, let us say, x=0. Substituting x=0 in the expression makes it $\frac{1}{1+0} + \frac{1}{1-0} = 1 + 1 = 2$. Substituting x=0 in the Choice A, makes it equal to 2. Hence, Choice A is a possible answer. When x=0, Choices B, C equal to 0. Choice D is equal to 0. Hence, you can eliminate Choices B, C, and D. Choice E is 1, and can be eliminated. Solving the problem in a traditional way,

$$\frac{1}{1+x} + \frac{1}{1-x} = \frac{1.(1-x) + 1.(1+x)}{(1+x).(1-x)} = \frac{1-x+1+x}{1-x+x-x^2} = \frac{2}{1-x^2}.$$

8. **Best answer is B.** Substituting x=y=2 leads to simplification: $\frac{1}{1-2} + \frac{1}{1-2} = \frac{1}{-1} + \frac{1}{-1} = -1 - 1 = -2$

   When x=y=2, Choice A becomes $\frac{1}{1-2-2-2\times2} = \frac{1}{1-2-2-4} = \frac{1}{-7}$; hence, it can be eliminated.

   When x=y=2, Choice B becomes $\frac{2-2-2}{1-2-2+2\times2} = \frac{-2}{1-2-2+4} = \frac{-2}{1} = -2$; hence, it is a possible answer.

   When x=y=2, Choice C becomes $\frac{2-2-2}{1-2-2} = \frac{-2}{1-2-2} = \frac{-2}{-3} = \frac{2}{2} = 1$; hence, it can be eliminated.

   Similarly, Choice D can be eliminated. Choice E is equal to 1 and it can be eliminated.
   Solving the problem in a traditional way,

$$\frac{1}{1-x} + \frac{1}{1-y} = \frac{1.(1-y) + 1.(1+x)}{(1-x).(1-y)} = \frac{1-y+1-x}{1-x-y-xy} = \frac{2-x-y}{1-x-y+xy}.$$

9. **Best answer is E.** $\frac{3}{x} + \frac{x+1}{x+5} = \frac{3.(x+5)+x.(x+1)}{x.(x+5)} = \frac{3x+15+x^2+x}{x.(x+5)} = \frac{x^2+4x+15}{x(x+5)}$.

10. **Best answer is A.** $\frac{x^2-1}{x+1} = \frac{(x-1).(x+1)}{(x+1)} = x - 1$.

Because numerator has x^2 and denominator has x, the answer should have x in the numerator. Hence, choices C and D can be eliminated. Similarly, numerator has -1 and denominator has 1 (and $\frac{-1}{1}$=-1), hence, Choice B can be eliminated. Choice E equals 1 and can be eliminated. Choice A makes logical sense.

11. **Best answer is D.**

$$\frac{X^2 - 9}{x + 3} \div \frac{x - 5}{x^2 - 25} = \frac{x^2 - 9}{x + 3} \cdot \frac{x^2 - 25}{x - 5} = \frac{(x + 3).(x - 3)}{x + 3} \cdot \frac{(x + 5).(x - 5)}{x - 5}$$

$$= (x - 3)(x + 5)$$

$$= x^2 - 3x + 5x - 15 = x^2 + 2x - 15.$$

3. Ratios

Ratios express relationships between two quantities. If there are 20 boys and 35 girls in a class, the ratio of boys to the girls in the class is 20:35. Just like fractions, ratios can be expressed in simple terms by dividing or multiplying them with the same number. The ratio 20:35 is same as 4×5:7×5 = 4:7. Ratios are used to express proportions. Two important concepts are direct proportion and indirect proportion.

Direct Proportion

Two quantities are said to be in direct proportion if

- The ratio of the two quantities is constant. i.e., $\frac{X}{Y}$ = constant.
  - The perimeter (P) and the radius (R) of a circle are in direct proportion because $\frac{P}{R} = 2\pi$.
- Equivalent fractions can be used to solve problems involving direct proportion.
- Note that if X goes up, Y goes up and vice versa. If X goes down, so does Y.
- Note also that in some problems, X and Y^2 can be in direct proportion.
  - $\frac{X}{Y^2}$ = constant.
  - The area (A) and the square of the radius of a circle (R) are in direct proportion because $\frac{A}{R^2} = \pi$.

Indirect Proportion

Two quantities are said to be in indirect proportion if

- The product of the two quantities is constant. i.e., XY = constant.
- Note that if X goes up, Y goes down and vice versa.
- Note also that in some problems, X and Y^2 can be in indirect proportion.
  - $X Y^2$ = constant.

Review the example problems below:

Example 1: My graduating class of 48 had 16 girls. What is the ratio of boys to the girls?

Number of boys = 48-16=32. Number of girls = 16. Ratio of boys to girls = 32:16 = 2:1 (and not 1:2).

Example 2: If a recipe uses 3 grams of sugar and 12 grams of flour. How many grams of sugar are needed when mixing 18 grams of flour?

Sugar	Flour
3	12
X	18

$$\frac{3}{X} = \frac{12}{18}; 3 \times 18 = 12 \times X; X = \frac{3 \times 18}{12} = 4.5; \text{note that } 12 \times 4.5 = 3 \times 18 = 54.$$

Example 3: If 5 gallons of gas cost $12, how much do 8 gallons of gas cost?

Gallons	Price
5	$12
8	$X

$$\frac{5}{8} = \frac{12}{X}; 5 \times X = 12 \times 8; X = \$\frac{12 \times 8}{5} = \$19.2; \text{note that } 5 \times \$19.2 = 8 \times \$12 = \$96.$$

Example 4: There are twice as many oranges as apples. There are six times as many apples as mangoes. What is the ratio of oranges to mangoes?

Let us say, number of mangoes = M. Number of apples = 6M. Number of oranges = 2×number of apples = 2×6×M = 12M. Ratio of oranges to mangoes = 12M:M=12:1.

Example 5: If interior angles in a triangle are in the ratio of 1:2:3, what is the largest angle?

Method 1: For this type of a problem, first find out the sum of the parts. In this case, sum of the parts = 1+2+3=6. Sum of interior angles in a triangle = 180 degrees. The largest angle is the one that is with proportion "3".

$$\text{Largest angle} = \frac{3}{\text{Sum of the parts}} \times 180 = \frac{3}{6} \times 180 = 3 \times 30 = 90.$$

 Note that it is a right triangle. This triangle is a 30, 60 and 90 degree triangle.

Method 2:

Angle	Total
3	6
Largest Angle	180

The largest angle $= \frac{3}{6} \times 180 = 3 \times 30 = 90$ degrees.

Method 3: When the ratios are given, assume the angles as X, 2X and 3X (because they are in the ratio 1:2:3). The sum of the angles = X+2X+3X = 6X = 180 (the sum of angles in a triangle is equal to 180 degrees).

Solving for X, X = $\frac{180}{6}$ = 30 degrees. The largest angle is 3X = 3×30 = 90 degrees.

SGK's Short Cut: Ratios 1+2 =3. In a right triangle, the sum of two complementary angles = 90 degrees = third angle. Hence, if the angles are in the ratio 1:2:3, it is a right triangle and the largest angle is equal to 90 degrees. For problems involving angles in a triangle, when two ratios equal to the third ratio, it is a right triangle.

Angles are in the Ratio	Property	Angles in a triangle equal to
1:2:3	1+2=3	30, 60 & 90
1:1:2	1+1=2	45, 45 & 90
2:3:5	2+3=5	36, 54 & 90
1:4:5	1+4=5	18, 72 & 90
1:3:4	1+3=4	22.5, 67.5 & 90
1:8:9	1+8=9	10, 80 & 90

Example 6: At a picnic, there are 3 men for every 4 women. What is the ratio of total attendees to the men at the picnic?

Method 1: Let a=men and b=women at the picnic and because $\frac{a}{b} = \frac{3}{4}$, we can assume that a = 3k (where "k" is a constant) and b = 4k. Note that, now, $\frac{a}{b} = \frac{3k}{4k} = \frac{3}{4}$. Now, calculate (a+b) in terms of "k". a+b = 3k+4k = 7k. $\frac{a+b}{a} = \frac{7k}{3k} = \frac{7}{3}$.

SGK's Short Cut: In this type of a problem, it is alright to assume a = 3 and b = 4. (a+b) = 3 + 4 = 7. Hence, $\frac{(a+b)}{a} = \frac{3+4}{3} = \frac{7}{3}$.

Example 7: At a carnival, there are men, women and children. The ratio of men to women to children is 3:5:11, what is the ratio of men to the total in attendance?

SGK's Short Cut: You can assume that a=3, b=5 and c=11. $\frac{a}{a+b+c} = \frac{3}{3+5+11} = \frac{3}{19}$.

Example 8: A fruit shop sells apples, oranges and mangoes. At the end of the day, the owner finds that twice the number of apples sold = three times the number of oranges sold = five times the number of mangoes sold; what is the ratio of the total fruits sold to the number of oranges sold?

SGK's Short Cut: Assume, a=apples, b=oranges and c=mangoes. 2a=3b=5c. In this problem, you are <u>not</u> able to say a=2, b=3 and c=5. If you do, you readily see that 2a=4, 3b=9 and 5c=25 (4 ≠ 9 ≠ 25). We need to find a new way.

Note that 2, 3 and 5 are multiples of a, b and c. The product of 2, 3, and 5 is equal to $2 \times 3 \times 5 = 30$. Assume, 2a, 3b and 5c are equal to the product $2 \times 3 \times 5 = 30$. Because 30 is divisible by 2, 3, and 5, it is easy to solve for a, b and c. Solving the equation 2a = 3b = 5c = 30, a = 3×5=15, b=2×5=10 and c=2×3=6.

$$\frac{a+b+c}{b} = \frac{15+10+6}{10} = \frac{31}{10}.$$

<u>Example 9</u>: X, Y and Z are angles in a triangle. If X:Y = 2:3 and Y:Z = 3:5, what is Z?

The sum of angles in a triangle is equal to 180. Hence, X+Y+Z=180. In addition, it is given that $\frac{X}{Y} = \frac{2}{3}$ and $\frac{Y}{Z} = \frac{3}{5}$. Y is the common element. Hence, express X and Z in terms of Y. $X = \frac{2}{3}Y$ and $Z = \frac{5}{3}Y$. $\frac{2}{3}Y + Y + \frac{5}{3}Y = 180$. Solving for Y, $\frac{10}{3}Y = 180$ and $Y = 18 \times 3 = 54$ degrees. X = 36 and Z = 90 degrees.

<u>Example 10</u>: Two stations A and B are 500 miles apart. A train leaves station A at 9 a.m. and is traveling at a speed of 100 miles per hour. Another train leaves station B at 9 a.m. and is traveling at a speed of 150 miles per hour. At what time do the two trains cross each other?

Let us assume that the two trains meet after time "t" hours. First train will travel a distance of 100×t miles. Second train will travel a distance of 150×t miles. Total distance traveled by both trains = 500 = 100t+150t=250t. Solving for t, t=$\frac{500}{250}$=2 hours. So, both trains meet after 2 hours i.e., 11 a.m.

SGK's Short Cut: $t = \frac{500}{100+150} = \frac{500}{250} = 2.$

<u>Example 11</u>: Richard went from Chicago to St. Louis at an average speed of 60 miles per an hour. On the return trip, he averaged 70 miles per hour. What was his average speed for the entire trip?

The answer is not 65. Let us assume that the distance between Chicago and St. Louis is "x" miles. Total distance traveled for the round trip is x+x=2x.

Time taken by Richard to go from Chicago to St. Louis is $\frac{x}{60}$.

Time taken for the return trip = $\frac{x}{70}$.

Total time taken for the round trip = $\frac{x}{60} + \frac{x}{70} = \left(\frac{70+60}{4200}\right)x = \frac{130}{4200}x.$

Average speed=$\frac{\text{Total distance traveled}}{\text{Total time taken for the trip}} = \frac{2x}{x\frac{130}{4200}} = 2 \times \frac{4200}{130} = 64.62$ miles per hour.

SGK's Short Cut: For a problem where the distance traveled in either direction is the same, average speed can be obtained by using the formula below:

$$\frac{2}{V} = \frac{1}{V_1} + \frac{1}{V_2}$$ where V is the average speed, V_1 and V_2 are speeds in the respective directions.

Solving for V, $V = \frac{2 \times V_1 \times V_2}{V_1 + V_2}.$

The above formula is an important one to note. It represents harmonic mean. Harmonic mean is one that is calculated by the reciprocals. Harmonic mean is always less than (or equal to) the arithmetic mean. In the previous example, the arithmetic mean is 65 mph. Note that the harmonic mean (64.62) is less than 65. You may be able to use this property to eliminate choices.

<u>Example 12</u>: If Randy completes a job in 4 days and Rachel completes the job in 6 days, how long do they take to complete the job working together?

Randy completes $\frac{1}{4}$ work in a day. Rachel completes $\frac{1}{6}$ work in a day. Both of them together complete $\frac{1}{4} + \frac{1}{6} = \frac{2+3}{12} = \frac{5}{12}$ work in a day. So, together they will take $\frac{12}{5} = 2.4$ days to complete the job.

SGK's Short Cut: If there are only two workers, you may calculate the combined rate using the formula:

Combined rate $= \frac{r_1 \times r_2}{r_1 + r_2}$ where r_1 and r_2 are the individual rates; $= \frac{1}{2} \times$ harmonic mean.

Using the formula, $\frac{4 \times 6}{10} = \frac{24}{10} = 2.4$ days.

Combined rate $\geq \frac{1}{2}$ (faster of the two rates) $\geq \frac{1}{2} \times 4 = 2$.

In the example, arithmetic mean is 5 days. $\frac{1}{2}$ of arithmetic mean is 2.5 days. Hence, our answer should be less than 2.5 days. Hence, $2 \leq$ combined rate ≤ 2.5. You may use this property to eliminate choices.

Properties of a Combined Rate (when only two objects are involved)

- Combined rate $\geq \frac{1}{2}$ (faster of the two rates).
- Combined rate $\leq \frac{1}{2}$ (arithmetic mean of the two rates).

r_1	r_2	Faster of the Two Rates	Arithmetic Mean	Combined Rate	Combined Rate Using the Formula
10	20	10	15	$5 \leq r \leq 7.5$	$\frac{10 \times 20}{10 + 20} = \frac{200}{30} = 6.67$
30	50	30	40	$15 \leq r \leq 20$	$\frac{30 \times 50}{30 + 50} = \frac{1500}{80} = 18.7$
40	50	40	45	$20 \leq r \leq 22.5$	$\frac{40 \times 50}{40 + 50} = \frac{2000}{90} = 22.2$
20	60	20	40	$10 \leq r \leq 20$	$\frac{20 \times 60}{20 + 60} = \frac{1200}{80} = 15$
80	120	80	100	$40 \leq r \leq 50$	$\frac{80 \times 120}{80 + 120} = \frac{9600}{200} = 48$

Example 13: I have two faucets at home. If first faucet takes 60 minutes to fill my tub and the second faucet takes 90 minutes to fill the tub, how long does it take to fill the tub when both faucets are open?

Faucet A fills $\frac{1}{60}$ in one minute and faucet B fills $\frac{1}{90}$ in one minute. Both faucets together will fill $\frac{1}{60} + \frac{1}{90} = \frac{3+2}{180} = \frac{5}{180}$ in one minute. So, both faucets take $\frac{180}{5} = 36$ minutes to fill the tub.

SGK's Short Cut:

Similar to the previous example, you may calculate the combined rate using the formula:

Combined rate $= \frac{r_1 \times r_2}{r_1 + r_2}$ where r_1 and r_2 are individual rates; $= \frac{1}{2} \times$ harmonic mean.

Using the formula, combined rate $= \frac{60 \times 90}{60 + 90} = \frac{5400}{150} = 36$ minutes.

Also, $\frac{1}{2} \times 60 \leq$ combined rate $\leq \frac{1}{2} \times (60 + 90)$.

$30 \leq$ combined rate ≤ 37.5.

Example 14: I have faucet A that takes 60 minutes to fill a tub and faucet B that takes 90 minutes to fill the tub. I also have a drain that takes 120 minutes to empty the tub. One day, I opened both faucets to fill the tub. I did not realize my son accidently left the drain open. How long does it take to fill the tub when both faucets and the drain are open?

In one minute, $\frac{1}{60} + \frac{1}{90} - \frac{1}{120} = \frac{6+4-3}{360} = \frac{7}{360}$ of tub is filled.

Hence, it takes $\frac{360}{7} = 51.4$ minutes to fill the tub.

Example 15: At a Summer camp, a ration lasts 16 days for 2 men and 3 children, how long does the ration last for 3 men and 8 children?

Ration	Men	Children
16	2	3
X	3	8

Because there are more men, the ration will last less number of days. Multiply 16 with $\frac{2}{3}$.

Similarly there are more children, the ration will last less number of days. Multiply 16 with $\frac{3}{8}$.

The ration will last $16 \times \frac{2}{3} \times \frac{3}{8} = 4$ days.

Practice Problems

1. Angles in a right isosceles triangle are in the ratio of
    A. 1:2:7   B. 1:1:5   C. 1:1:1   D. 1:2:3   E. 1:1:2

2. Weight of an object on earth is 10 lb. Weight of an object on a planet is proportional to the gravitational force of the planet. What is the weight of the same object on Moon if the gravitational force on Moon is 6 times less that the gravitational force on earth?
    A. 60 lb   B. 6 lb   C. 600 lb   D. 1.67 lb   E. 0.6 lb

3. A student in my class observed angles in a triangle to be in the ratio of 5:7:12. What type of a triangle is it?
    A. Scalene   B. Isosceles   C. Equilateral   D. Right Triangle   E. Right
        Isosceles Triangle

4. The force between two objects is indirectly proportional to square of the distance between the objects. If the force between two objects when they are 5 feet apart is 100 units, what is the force between the objects when they are 50 feet apart?
    A. 10 units   B. 1000 units   C. 1 unit   D. 100 units   E.10000 units

5. Two angles on a straight line are in the ratio of 3:7, what is the larger angle equal to?
    A. 54 degrees B. 30 degrees   C. 90 degrees   D. 45 degrees   E. 126 degrees

6. If my college baseball team wins half of the games in a season and draws $\frac{1}{5}$ of the games, what is the win-loss ratio of the team for the season?
    A. $\frac{5}{9}$ B. $\frac{5}{3}$ C. $\frac{1}{2}$ D. $\frac{2}{5}$ E. $\frac{4}{5}$

7. α, β and γ represent angles in a triangle. If α:β = 1:2 and β:γ = 2:5, what is α?
    A. 90   B. 45   C. 22.5   D. 30   E. 60

8. For every 4 students who passed a test, there are 7 students who failed the test. What fraction of the students passed the test?
    A. $\frac{7}{4}$ B. $\frac{11}{4}$ C. $\frac{4}{11}$ D. $\frac{4}{3}$ E. $\frac{3}{4}$

9. A shop is selling red, white and blue t-shirts. If the number of red shirts sold = three times the white shirts sold=five times the blue shirts sold. What fraction of the shirts sold are red?
    A. $\frac{15}{23}$ B. $\frac{5}{23}$ C. $\frac{1}{23}$ D. $\frac{1}{9}$ E. $\frac{3}{23}$

10. A shop is selling red, white and blue t-shirts. If the number of red shirts sold to the white shirts is 1:3 and the ratio of white shirts to the blue shirts is 2:5, what fraction of the shirts sold are red?
    A. $\frac{15}{23}$ B. $\frac{5}{23}$ C. $\frac{1}{23}$ D. $\frac{1}{9}$ E. $\frac{2}{23}$

11. At a parade, 3 out of every 7 children are holding a flag. What is the ratio of the total children at the parade to the children not holding a flag?
    A. $\frac{7}{4}$ B. $\frac{7}{3}$ C. $\frac{4}{3}$ D. $\frac{3}{10}$ E. $\frac{7}{10}$

12. At a party, men are wearing red shirts and women are wearing white shirts. If there are 5 red shirts for every 7 white shirts, what could be one value of the total attendees at the party?

   A. 5   B. 7   C. 24   D. 30   E. 40

13. A train leaves a station at 9 a.m. It is traveling at a constant rate of 50 mph. After 3 hours, another train leaves the station (in the same direction as the first train) at 100 mph. At what time do the two trains meet?

   A. 12 noon   B. 1 p.m.   C. 2 p.m.   D. 3 p.m.   E. 4 p.m.

14. Two trains leave a station in opposite directions with an average speed of 40 mph and 60 mph respectively. How far are they apart in 3 hours?

   A. 120 miles   B. 180 miles   C. 240 miles   D. 300 miles   E. 360 miles

15. If Ben finishes a job in 8 days, Brady in 12 days and Sydney in 18 days, how long does it take for them to finish the job working together?

   A. 12 days   B. 4 days   C. 6 days   D. 2 days   E. 3.8 days

16. Faucet A takes 4 hours to fill a tub and faucet B takes 9 hours to fill the tub. Faucet A was used to fill the tub for 2 hours. Then, both faucets were turned on to fill the tub. How long did it take to fill the tub (in total)?

   A. 2 hours   B. 3.38 hours   C. 2.77 hours   D. 3 hours   E. 3.25 hours

17. 2 men, 3 women and 8 children take 24 days to finish a job. How long does it take for 3 men, 2 women and 16 children to complete the same job?

   A. 24 days   B. 8 days   C. 12 days   D. 6 days   E. 4 days

Solutions to Practice Problems

1. ***Best answer is E***. A right triangle has a right angle = 90 degrees. Because it is an isosceles triangle the remaining two angles must be equal. Hence, each of the two remaining angles should be equal to 45 degrees. All three angles add up to 180 degrees (90+45+45=180). The ratio is 45:45:90=1:1:2. *SGK's Short Cut:* You may also note that in the ratio 1:1:2, 1+1=2, i.e., the sum of two parts equals the third which also means it is a right triangle because the sum of complementary angles is equal to 90 degrees. Two of the angles are equal because the ratio is 1:1.

2. ***Best answer is D***. The weight on the Moon will be 6 times less than the weight on the earth. Hence, the weight on the Moon will be = $10 \times \frac{1}{6}$ = 1.67lb.

3. ***Best answer is D***. Let the angles be 5X, 7X and 12X (in the ratio 5:7:12). Because the sum of angles in triangle equals 180, 5X+7X+12X = 180 degrees. Solving for X, 24X=180 and X = $\frac{180}{24}$ = 7.5 degrees. Hence, the angles are 37.5, 52.5 and 90 degrees. The triangle is a right triangle. *SGK's Short Cut:* Note also that 5+7=12. The sum of two parts equal to the third part. If you recognize this, you do not even need to solve for X. Hence, it is a right triangle because complementary angles add up to 90 degrees.

4. **Best answer is C.** Force × Distance × Distance = Constant. 100 × 5 × 5 = F × 50 × 50. Solving for F, F = 1 unit.

5. **Best answer is E.** <u>Method 1:</u> Angles on a straight line add up to 180 degrees. Let the angles be 3X and 7X. 3X + 7X = 180; 10X=180; or X = 18. We need the larger angle. 7X = 7×18=126 degrees. *SGK's Short Cut:* All choices except Choice E can be eliminated right away. Two angles add up to 180. Half of 180 is 90 degrees. Because the question is about the larger angle, note that the larger angle should be greater than half of 180 i.e., greater than 90 degrees.

6. **Best answer is B.** <u>Method 1:</u> Win + Draw + Loss = 1; $\frac{1}{2} + \frac{1}{5}$ + Loss = 1 or Fraction of games that were lost = 1 $-(\frac{1}{2} + \frac{1}{5})$ = $\frac{3}{10}$. Win-Loss Ratio = $\frac{1/2}{3/10}$=$\frac{1}{2} \times \frac{10}{3} = \frac{5}{3}$. *SGK's Short Cut:* Because half the games are won, win loss ratio should be greater than 1. All choices except B can be eliminated because Choices A, C, D and E are less than 1.

7. **Best answer is C.** <u>Method 1:</u> α:β:γ=1:2:5. We are able to combine the ratio because β=2 in both the ratios. Let us assume α = "a", β = 2a and γ = 5a. α+β+γ = a+2a+5a=8a. However, we know that angles in a triangle add up to 180 degrees. Hence, 8a = 180 or a = 22.5 degrees. Hence, α = 22.5 degrees.

   <u>Method 2:</u> Let us assume α = 10 degrees as your starting guess. β = 2α = 2x10=20 degrees. γ = 20×$\frac{5}{2}$=50 degrees. α+β+γ=10+20+50=80 degrees. However, we know that α, β and γ should add up to 180 degrees. Hence, make it a simple ratio where $\frac{10}{α} = \frac{80}{180}$ and solving for α = $\frac{180}{8}$ = 22.5 degrees.

8. **Best answer is C.** Simply assume a = 4 and b = 7. $\frac{a}{a+b} = \frac{4}{4+7} = \frac{4}{11}$.

9. **Best answer is A.** *SGK's Short Cut:* Let, R = the number of Red shirts, W = the number of white shirts and B = the number of blue shirts. First we need to find out the ratio of R, W and B. <u>The ratio is not 1:3:5.</u> Take the multiples of R, W and B. In this case, 1, 3 and 5. 1x3x5=15. Let us say, R=3W=5B=15. Solving for R, W and B yields R = 15, W=5 and B=3. R:W:B=15:5:3.

$$\frac{R}{R+W+B} = \frac{15}{15+5+3} = \frac{15}{23}.$$

10. **Best answer is E.** Let, R = the number of Red shirts, W = the number of white shirts and B = the number of blue shirts. Note that W is not the same in both the ratios and hence, we cannot combine them. $\frac{R}{W} = \frac{1}{3} = \frac{2}{6}$; $\frac{W}{B} = \frac{2}{5} = \frac{6}{15}$; now that y = 6 in both the ratios, we can combine them.

R: W: B = 2: 6: 15. $\frac{R}{R+W+B} = \frac{2}{2+6+15} = \frac{2}{23}.$

11. **Best answer is A.** Assume a=the number of children holding the flag and b=the total number of children at the parade. $\frac{b}{b-a} = \frac{7}{7-3} = \frac{7}{4}.$

12. **Best answer is C.** R+W=5+7=12. Hence, R+W should be a multiple of 12. Only <u>Choice C</u> is possible.

13. **Best answer is D.** <u>Method 1:</u> Let "t" is the time it takes both trains to meet. 50t=100(t-3). Solving for t, t=6hours. Hence, both trains meet after 9 a.m. + 6 = 3 p.m.. *SGK's Short Cut:* In the first 3 hours, train 1 has traveled 50×3=150 miles. Second train is closing the gap by 100-50=50mph. Hence, it takes $\frac{150}{50}$=3 hours for it to close the gap. Hence, both trains meet at 3 p.m.

14. **Best answer is D.** *SGK's Short Cut:* Because they are going in opposite directions, after one hour both trains will be 40+60=100 miles apart. After 3 hours, they will be 100×3=300 miles apart.

15. **Best answer is E.** After one day, together they will finish $\frac{1}{8} + \frac{1}{12} + \frac{1}{18} = 0.2639$. Hence, it will take $\frac{1}{0.2639} = 3.8$ days to complete the job.

16. **Best answer is B.** After two hours, faucet A has filled $\frac{1}{2}$ of the tub. Together, both faucets will take $\frac{4×9}{13} = \frac{36}{13} = 2.769$ hours to fill the tub. However, they only need to fill half the tub. Hence, they will take 1.384 hours to fill half the tub. Overall, it takes 2+1.384=3.384 hours to fill the tub.

17. **Best answer is C.** Multiply 24 with the ratios $\frac{2}{3}, \frac{3}{2}$ and $\frac{8}{16}$. i.e., $24 \times \frac{2}{3} \times \frac{3}{2} \times \frac{8}{16} = 12$ days.

4. Percents

What is a "Percent"?

"Per" means division. "Cent" means century or hundred.

"Percent" means division with hundred or a ratio with hundred. A Percent can be expressed as *a decimal number* or *a fraction* or *a ratio*.

For example, $10\% = \frac{10}{100} = \frac{1}{10} = 0.1$.

$20\% = \frac{20}{100} = \frac{1}{5} = 0.2$.

$45\% = \frac{45}{100} = \frac{9}{20} = 0.45$.

"Percent" problems are related to "Ratio", "Fraction" or "Proportion" problems.

Use the following short cuts for Percent problems.

Percent	Multiply by Fraction	Multiply by Decimal Number
0.1%	$\frac{0.1}{100}$ or $\frac{1}{1000}$	0.001
1%	$\frac{1}{100}$	0.01
5%	$\frac{5}{100}$ or $\frac{1}{20}$	0.05
10%	$\frac{1}{10}$	0.1
20%	$\frac{1}{5}$	0.2
25%	$\frac{1}{4}$	0.25
30%	$\frac{3}{10}$	0.3
$33\frac{1}{3}\%$	$\frac{1}{3}$	0.33
50%	$\frac{1}{2}$	0.5
75%	$\frac{3}{4}$	0.75
80%	$\frac{4}{5}$	0.8
90%	$\frac{9}{10}$	0.9
100%	1	1
110%	$\frac{11}{10}$	1.1
120%	$\frac{12}{10}$	1.2

SGK's Short Cut:

- Note that 10% < 20% < 30% < 40% < 50% ...etc.. < 80% < 90% < 100%.
- 100% means "All of it". 100% of any number is equal to the number itself.
- Use 10% as a reference. 20% is twice that of 10%. 30% is 3 times that of 10%. 40% is 4 times that of 10%. You may use proportions to solve the problems involving percents.
  - For example, 10% of 60 is 6. Hence, 20% of 60 = 2×6=12. 30% of 60=3×6=18. 40% of 60=4×6=24.

Item	Short Cut
To find out 1%	• Remove two zeros from the end or add two decimal places • 1% of 1200 is 12. • 1% of 1256 is 12.56.
To find out 10%	• Remove one zero from the end or add one decimal from the end. You may use 10% as a reference. • 10% of 1200 is 120; 10% of 150 is 15. • 10% of 1256 is 125.6; 10% of 156 is 15.6.
To find out 25%	• Note that 25% is one quarter; hence divide by 4 or multiply by 0.25. • 25% of 120 is 30; 25% of 1200 is 300.
To find out 50%	• 50% means $\frac{1}{2}$; hence, multiply by $\frac{1}{2}$ or 0.5 or divide by 2. • 50% of 240 is 120.
To find out selling price at 10% discount	• 10% discount means 10% off of the original price. 100% means full price. So, the selling price = 100% - 10% = 90%. Hence, multiply by 0.9.
To find out selling price at 20% discount	• 20% discount means 20% off of the original price. 100% means full price. So, the selling price = 100% - 20% = 80%. Hence, multiply by 0.8.
To find out selling price at 30% discount	• Multiply by 0.7.
To find out 10% increase	• 10% increase means 10% more than the original price. 100% means full price. So, the selling price = 100% + 10% = 110%. Hence, multiply by 1.1.
To find out 20% increase	• Multiply by 1.2.
To find out 50% increase	• Multiply by 1.5.
To find out increase of 10% and 20%	• Multiply by 1.1 and 1.2 = 1.32; which means 32% increase • There is another simple way. 10% + 20% + (10% of 20 = 2%) = 32%.
To find out increase of 20% and 40%	• Multiply by 1.2 and 1.4 = 1.68 or 68% increase. • OR simply, 20% + 40% + (20% of 40 = 8%) = 68%.
To apply two successive discounts of 10% and 20%	• Multiply 0.9 and 0.8 and subtract from 1. • i.e., 1-(0.8x0.9) = 1 – 0.72 = 0.28 or 28%. • Two successive discounts will always be less than two of them combined. i.e., in the above example, less than 10%+20% = 30%. • OR Simply, 10% + 20% - (10% of 20) = 10%+20%-2%=28%.
To apply two successive discounts of 20% and 30%	• Multiply 0.8 and 0.7 and subtract from 1. • i.e., 1-(0.8x0.7) = 1 – 0.56 = 0.44 or 44%. • OR Simply, 20% + 30% - (20% of 30) = 20%+30%-6%=44%.

You must learn the following four different types of percent problems.

Example 1: 10% of 150 fruits at a shop are spoiled at the end of the day. How many fruits are spoiled?

Use the following to translate the problem as:

What means "X". "Is" means "=". 10% means $\frac{10}{100}$. "Of" means "multiplication".

$X = \frac{10}{100} \times 150$; Solve for X, $X = \frac{150}{10} = 15$. Or simply, multiply 150 with 0.1 because 10% means 0.1.

Example 2: A basketball team lost 6 out of the 24 games in a season. What percent of the games are lost?

Once again, translate the problem as:

"What" means X. Percent means division by 100. "Of" means multiplication. "Is" means "=".

$\frac{X}{100} \times 24 = 6$; Solving for X, $X = 6 \times \frac{100}{24} = 25\%$.

Example 3: 20% at a party are children. If there are 30 children at the party, how many people are attending the party?

$30 = \frac{20}{100} X$; Solving for X, $X = 30 \times \frac{100}{20} = 150$.

Percent Change

Percent increase or decrease is always calculated based on the original quantity or the starting point.

Percent Change $= \frac{\text{Increase (or decrease)}}{\text{Original Quantity}} \times 100$.

Example 4: Stock prices of company XYZ in 2010 and 2011 are \$55 & \$60 respectively. What is the percent change in company XYZ's stock price?

Percent Change $= \frac{\text{Increase in Stock Price}}{\text{Starting Value}} \times 100 = \frac{5}{55} \times 100 = 9\frac{1}{11}\%$.

Example 5: In 2010, the S&P index increased by 25%. By what percent, does the S&P index fall in 2011 to equal the value at the beginning of 2010?

Because we do not know what the value of the S&P index is at the beginning of 2010, we can assume it to be 100. Assuming the value to be 100 makes the problem easy. 25% from 100 = 100 + 25% of 100 = 100 + 25 = 125. To return to the original value (which is 100), S&P index has to decrease by 125-100 = 25 points. Percent decrease $= \frac{25}{125} \times 100 = 20\%$; not same as 25%.

SGK's Short Cut:

25% increase means 1.25 or $\frac{5}{4}$. To find out % decrease needed to return to the original value, multiply by the reciprocal. i.e., $\frac{4}{5}$. Note that $\frac{4}{5} = 0.8$ or equals to 20% decrease.

Practice Problems

1. My son answered 15% of the questions incorrectly in a test consisting of 240 questions. How many questions did he answer incorrectly?
    A. 24   B. 36   C. 400   D. 3.6   E. 0.36

2. At a stadium, 72 seats out 360 are empty. What % of the stadium is not filled?
    A. 120%   B. 20%   C. 5%   D. 10%   E. 2%

3. A treasure is divided between my son and my daughter in the ratio of 75% and 25%. My daughter received $35. How big is the treasure?
    A. $70   B. $14000   C. $3.5   D. $3500   E. $140

4. You have been saving for a laptop that is listed at a price of $140. This week, the store is advertising a sale of 15%. How much do you need to save to buy the laptop?
    A. $125   B. $155   C. $120   D. $200   E. $119

5. A school has 45% boys and the rest girls. If there are 900 enrolled students in the school, how many more girls are in the school than boys?
    A. 105   B. 185   C. 80   D. 90   E. 9

6. If a stock price increased by 20% in 2001 and 30% in 2002. What is the overall percent increase?
    A. 50%   B. 56%   C. 24%   D. 60%   E. 90%

7. At a store, loyalty members receive 10% discount. For July 4th, the store is offering additional 20% discount. What is the overall discount received by the loyalty members?
    A. 30%   B. 28%   C. 50%   D. 21%   E. 15%

8. If there are 1200 students in a class and there are 240 more boys in the class than girls, what percent of the class are girls?
    A. 40%   B. 60%   C. 80%   D. 20%   E. 50%

9. My son draws two circles. Second circle has a radius 10% longer than the first circle. By what percent is the circumference of the second circle longer than the first one?
    A. 10%   B. 21%   C. 20%   D. π multiplied by 10%   F. 5%

10. My daughter draws two circles. Second circle has a radius 15% longer than the first circle. By what percent is the area of second circle greater than the first one?
    A. 15%   B. 30%   C. 20%   D. 32.25%   E. 225%

11. If area of an equilateral triangle is increased by 44%, what is the percent increase in the side of the triangle?
    A. 10%   B. 20%   C. 30%   D. 44%   E. 15%

12. A rectangular parcel of land in Texas is sold per sq.ft. If two sides of a rectangular plot are increased by 20% and 30% respectively, what is the percent increase in the price of the plot?
    A. 50%   B. 60%   C. 30%   D. 56%   E. 90%

13. The cost of construction is proportional to the area of the plot. If the length of a rectangle is increased by 25% and width is reduced by 20%, what is the percent change in the construction cost?
    A. 20%   B. 5%   C. 50%   D. 45%   E. No change

14. If a stock price changes from $120 to $144, what is the percent increase in the stock price?
    A. 24%   B. 10%   C. 200%   D. 2%   E. 20%

Solutions to Practice Problems

1. ***Best answer is B***. $X = \frac{15}{100} \times 240 = 36$. 10% of 240 is 24. Hence, the answer should be more than 24. Choices A, D and E can be eliminated. Choice C is too big.

2. **Best answer is B.** $72 = \frac{X}{100} \times 360$; $X = 20$. 10% of 360 is 36. Hence, using ratios or proportions, 72 must be 20% of 360. Choices C, D and E are too low and Choice A is too high.

3. **Best answer is E.** 25% is same as one quarter or $\frac{1}{4}$. Hence, we are looking for 35×4=140. Choices B and D are too high and C is too low. 35 is 50% of 70 and hence, Choice A can be eliminated.

4. **Best answer is E.** If there is a sale or a discount, you will pay a lower price than the original price of $140. Hence, Choices B and D can be eliminated. With 15% discount, you will only have to pay 100-15=85% of the original price. Hence, you need to save $140×0.85=$119.

5. **Best answer is D.** *SGK's Short Cut:* Here is another time saving tip. Note that in this type of a problem, you only need to find the difference. No need to find out the number of boys and girls individually. If there are 45% boys in the school, 55% are girls. The difference between girls and boys = 55%-45%=10%. Hence, we need to find out 10% of 900= 900×0.1=90.

6. **Best answer is B.** Choice D and E are too high and Choice C is too low. Choice A is not the answer.

   <u>Method 1:</u> Let us say the initial price is $100. 20% increase means $120. Now apply 30% increase = $120×1.3=$156 which equates to an overall increase of 56%.

   SGK's Short Cut: We also talked about cumulative increase. In this case, cumulative increase is 20% + 30% + (20% of 30) = 20% + 30% + 6% = 56%.

7. **Best answer is B.** Assume the original price is $100. The purchase price is $100×0.9×0.8=$72. Cumulative discount $= \frac{\$100-\$72}{\$100} = 28\%$. Note that two successive discounts will not add up to 10% + 20%=30%.

   SGK's Short Cut: Cumulative discount = 10%+20%-(10% of 20)=30%-2%=28%.

8. **Best answer is A.** 1200-240=960. $\frac{960}{2} = 480$. Hence, there are 480 girls and 480+240=720 boys in the class. Percentage of girls $= \frac{480}{1200} \times 100 = 40\%$.

   SGK's Short Cut: $\frac{240}{1200} = 20\%$. % of boys $= 50\% + \frac{20\%}{2} = 50\% + 10\% = 60\%$. % of girls=50%-$\frac{20\%}{2}$=50%-10%=40%. In other words, you need to find two percents such that they add up to 100% and the difference between the two percents is 20%. i.e. 60% and 40%.

9. **Best answer is A.** Let us say, the radius of the first circle is 10. The circumference is 20π. Radius of the second circle is 10% longer, the new radius is 11. New circumference is 22π. The percent change is circumference is $\frac{22\pi - 20\pi}{20\pi} = \frac{2\pi}{20\pi} = \frac{1}{10} = 10\%$.

   SGK's Short Cut: Radius and circumference increase by the same percent. Pick 10%.

10. **Best answer is D.** <u>Method 1:</u> 1.15×1.15=1.3225. Hence, the area will increase by 1.3225-1=0.3225=32.25%.

   SGK's Short Cut: Increase in Area = 15% + 15%+(15% of 15)=30%+2.25%=32.25%.

11. **Best answer is B.** Assume the side of the triangle is X. Note that the area is proportional to the square of the side. Area is increased by 44%. In other words, the problem is saying two successive increases equal to 44%. The percent increase in the side should be equal to 20% because 1.2×1.2=1.44 or 44% increase. i.e., $x^2 = 1.44$ and $x = 1.2$ meaning 20% increase.

12. **Best answer is D.** <u>Method 1:</u> Multiply 1.20 and 1.30 to get 1.56. Hence, the new area will be 56% more than the previous area. Hence, the percent increase in price = 56%.

   SGK's Short Cut: Change in Area = 20%+30%+(20% of 30)=50%+6%=56%.

13. **Best answer is E.** <u>Method 1:</u> Let us assume the sides are 100 and 200. The new sides are 125 and 160. Note that both the rectangles have the same area of 20000 sq. units. The area is unchanged. Hence, there is no change in the construction cost.

   SGK's Short Cut: New area = 1.25×0.8=1. Hence, there is no change in the area. Hence, there is no change in the construction cost.

14. **Best answer is E.** Increase in stock price = $144-120=$24. The percent increase $=\frac{24}{120} \times 100 = 20\%$.

5. Simple and Weighted Average

Simple Average or Arithmetic Mean or Mean

The average of a series of numbers is defined as the sum of the numbers divided by number of values in the series. For example, the average of the numbers 10, 16, 18 and 6 = $\frac{10+16+18+6}{4} = \frac{50}{4} = 12.5$.

Properties of an Average

- The average of two numbers is always the mid-point. Note that 28-14 = 14 and 42-28 = 14. In other words, the average of two numbers is equidistant from the two numbers.
- For a series of more than two numbers, the sum of the total deviations from the mean is always zero.
    - For example, average of 10, 16, 18 and 6 = 12.5. Deviations (or difference) from the mean are:
        - 10-12.5 = -2.5
        - 16-12.5 = 3.5
        - 18-12.5 = 5.5
        - 6-12.5 = -6.5.
        - Note that -2.5+3.5+5.5-6.5 = 0
- It helps to understand the impact of introducing a new number into a series. See below:
    - If a class scored an average of 80 points in a test. What happens to the average score if a new student joined the class and he/she scored
        - 90 in the test. Because 90 > 80, the class average will go up.
        - 70 in the test. Because 70 < 80, the class average will go down.
        - 80 in the test. Because it is the same number as the average, the class average will remain the same.
- 9 students in a class average 80 in a test. If new student joins the class and scores a 90 in the test, by how much does the class average go up?
    - The new average can be calculated as $\frac{(9\times80+90)}{10} = \frac{720+90}{10} = \frac{810}{10} = 81$.
    - The new student scored 10 more points than the class average. Including the new student, there are now 10 students in the class. Hence, the class average will go up by $\frac{10}{10}$ = 1 point. Hence, the new average = 80+1 = 81.
- In a series of consecutive integers, or in a series of consecutive odd integers or in a series of consecutive even integers, the average is always is *the middle number*. This is a very important concept.
    - For example, in the series of integers 7, 8, 9, 10, 11, the average is 9. In a series of odd numbers 11, 13, 15, 17, 19, 21, 23, the average is 17. In a series of even numbers 22, 24, 26, 28, 30, 32, 34, the average is 28. As you note from the above examples, it is easy to identify middle number if there are odd number of values in a series.
    - If there are an even number of values in a consecutive series, find the middle two numbers and find an average for the two middle numbers. For example, in the series of integers, 7, 8,

9, 10, 11, 12, the average is $\frac{9+10}{2} = 9.5$. In the series of odd numbers, 11, 13, 15, 17, 19, 21, 23, 25, the average is $\frac{17+19}{2} = 18$. In the series of even numbers, 22, 24, 26, 28, 30, 32, 34, 36, the average is $\frac{28+30}{2} = 29$.

- Sum of the numbers in a series = Average x N; where N is the number of items in the series.
  - For example, if there are 20 students in a class and the class average is 30; the sum of all scores in the class = 20x30 = 600.
  - Similarly, in the series, 7, 8, 9, 10, 11, the sum of the numbers = 9×5 = 45.
  - The sum of 11, 13, 15, 17, 19, 21, 23 = 17×7 = 119.
  - The sum of 22, 24, 26, 28, 30, 32, 34 = 28×7 = 196.

Weighted Average

"Weighted" means some values are given more importance than the others. On the other hand, when calculating a simple average, all values are given equal importance (or equal weight).

The weighted average is calculated as $\frac{W_1 X_1 + W_2 X_2}{W_1 + W_2}$; where W_1, W_2 are the weights and X_1 and X_2 are the individual averages. Refer to the following examples.

Example 1: In my son's class, there are 40 boys and 80 girls. For a fund raising event, each of the boys brings $4 and each of the girls brings $7, what is the average contribution of each of the students.

Note that there are 40+80 = 120 students in the class. Total contribution by the boys = 40×$4 = $160 The total contribution by the girls = 80×$7 = $560. The average contribution $= \frac{40 \times 4 + 80 \times 7}{40 + 80} = \frac{160 + 560}{120} = $6.

Properties of a Weighted Average

- The weighted average is always between the two individual averages. In the previous example, the weighted average ($6) is between $4 and $7.
- The weighted average is closer to the average with the higher weight. In the previous example, there are more girls in the class than boys. Hence, the weighted average ($6) is closer to $7 than to $4.
- The weighted average divides the individual averages in the inverse ratio of the weights. In the previous examples, the ratio of the weights = the ratio of boys to girls = 40 : 80 or 1:2; there are two girls to every boy. Note that $6-$4 = 2 and $7-$6 = 1. Hence, the weighted average $6 divides $4 and $7 in the ratio of 2:1 (inverse to 1:2).

Example 2: In a science experiment, 6 liters of 30% acid is mixed with 15 liters of 51% acid, what is the concentration of the resulting solution?

New concentration = $\frac{6 \times 30 + 15 \times 51}{6+15} = \frac{180+765}{21} = 45\%$. Note that there is more of 51% acid (15 liters), hence the new average is closer to 51%. Note that 6:15 = <u>2:5.</u> 45%-30% = 15% and 51%-45%=6%; 15%:6% = <u>5:2</u>.

Practice Problems

1. Five students in my class have scores of 82, 90, 16, 44 and 36 respectively. What is the average score?
   A. 10 B. 44 C. 53.6 D. 90 E. 58

2. The average rainfall per month in a year is 5 inches. What is the total rainfall in the year?
   A. 5 B. 60 C. 150 D. 48 E. 50

3. There numbers 1000, 1600 and "X" have an average of 1400. What is X?
   A. 1000 B. 1600 C. 1400 D. 1800 E. 4200

4. A museum is issuing tokens consisting of consecutive integers to its visitors. The last seven tokens issued average to 10. What is the smallest token issued?
   A. 13 B. 7 C. 10 D. 8 E. 14

5. A series consists of 13 consecutive integers. What is the largest number if the sum of the series is 0?
   A. 0 B. 13 C. 6 D. -6 E. 7

6. At a license station, tokens are issued consisting of consecutive odd integers. The last 11 tokens issued add up to 143. What is the 6th token issued?
   A. 7 B. 13 C. 11 D. 9 E. 15

7. An amusement park issues tokens consisting of consecutive even integers. The last 5 tokens issued average to 40. What is the latest token issued?
   A. 36 B. 38 C. 40 D. 42 E. 44

8. 5 Students in a class have an average score of 30. Perfect score in the test is 36. If John who scored a perfect score in the test joins the class, what is the new average of the class?
   A. 36 B. 30 C. 33 D. 31 E. 40

9. The ratio of boys to girls in a class is 2:3. If the average height of the boys is 70 inches and the average height of the girls is 66 inches, what is the average height of the class?
   A. 67.6 B. 74 C. 64 D. 68 E. 67

10. Two liters of 10% acid is mixed with eight liters of 60% acid. What percent acid is the resulting mixture?
    A. 10% B. 60% C. 35% D. 50% E. 80%

11. How much water should be added to 8 liters of 50% alcohol to make it 32% alcohol?
    A. 8 B. 10 C. 4 D. 4.5 E. 2

Solutions to Practice Problems

1. **Best answer is C.** The average score = $\frac{82+90+16+44+36}{5} = \frac{268}{5} = 53.6$.

2. **Best answer is B.** There are 12 months in a year. Total rainfall = 5×12=60 inches.

3. **Best answer is B.** 1000+1600+X = 3×1400=4200; solving for X, X=4200-1000-1600=1600.

4. **Best answer is B.** Note the numbers are _,_,_,10,_,_,_ because the average is always the middle number. Hence, the numbers are 7, 8, 9, 10, 11, 12, and 13. The smallest number is 7.

5. **Best answer is C.** The average = $\frac{0}{13}$ = 0; Hence, the middle number is 0. The largest number is 0+6=6.

6. **Best answer is B.** The middle number is the average of the numbers = $\frac{143}{11} = 13$. 6th number in the series is also the middle number. Hence, 6th term = 13.

7. **Best answer is E.** The average is the middle number = 40. The tokens are 36, 38, 40, 42 and 44. The latest token is 44.

8. **Best answer is D.** *SGK's Short Cut:* With John included, there are 6 students in the class. His score 36 is 6 more than the class average. Hence, $\frac{6}{6} = 1$. The new average is 1 more than the previous average = 30+1=31.

9. **Best answer is A.** Method 1: The new average = $\frac{2\times70+3\times66}{5} = \frac{140+198}{5} = \frac{338}{5} = 67.6$

   SGK's Short Cut: Weights are in the ratio 2:3. The range is 70-66=4. Divide 4 in the ratio of 2:3. The numbers will be 1.6 and 2.4. Hence, the weighted average = 66+1.6 or 70-2.4=67.6.

10. **Best answer is D.** Method 1: The weighted average = $\frac{2\times10+8\times60}{2+8} = \frac{20+480}{10} = \frac{500}{10} = 50\%$.
    SGK's Short Cut: The ratio of 10% acid to 60% acid is 2:8 = 1:4. So, divide the range, 60%-10%=50% in the ratio 1:4. Divide (60%-10%=50%) by $\frac{4}{1+4} = 50\% \times \frac{4}{5} = 40\%$. New solution is 10%+40% = 50% acid.

11. **Best answer is D.** Let us say, X liters of water is added. The new average is $\frac{X\times0+8\times50}{X+8} = \frac{400}{X+8} = 32$. Solving for X, X = 4.5.

6. Equations

Equations With One Variable

Equations with one variable involve one unknown quantity in the equation. Several techniques can be used to solve equations with one variable. These include substitution and separation of variables and constants.

Example 1: If 2X + 3 = 19, solve for X.

Step 1: Identify the terms involving variables and constants. In this case, "2X"; "3" and "19".
Step 2: Rearrange the terms so that all the variables are on one side of the equation and constants are on the other side of the equation.
By subtracting 3 from both sides, 2X + 3 -3 = 19-3; 2X = 16.
Step 3: Solve for X
Divide both sides by 2; $\frac{2X}{2} = \frac{16}{2}$; X = 8
Note that when you substitute X=8 in the equation; 2×8 + 3 = 16+3 = 19; Hence, X=8 satisfies the equation.

It is important to note <u>what</u> the problem is asking you to find out. Sometimes, you do not need to solve for X as noted in the following examples.

Example 2: If 4X + 6 = 20; what is 2X + 3?

Note 4X + 6 is twice that of 2X+3; hence 2X + 3 = $\frac{20}{2} = 10$.

Example 3: If 2X − 3 = 21; what is 4X?

From the above equation; 2X − 3 = 21 which means 2X = 21 + 3 = 24. 4X = 2×2X = 2×24 = 48.

Example 4: Seven years from now, Randy will be twice as old as he was two years ago. How old Randy be in two years?

Assume Randy is "X" years old now. Seven years from now; he will be X+7 years. Two years ago; he was X-2 years old.
Given X+7 = 2(X-2) = 2X-4; Solving for X; 2X-X = 7+4 or X = 11.
Trap: Note that the question is not asking for X; it is asking Randy's age two years from now. Which means, you need X+2 = 11+2 = 13.

Example 5: Three siblings are each one year apart in age. The sum of their ages is 207. How old is the youngest?

You can assume the numbers to be (X-1), X and (X+1). Add them together, you get X-1+X+X+1 = 3X. Hence, 3X = 207 (given); solving for X, X = $\frac{207}{3} = 69$. X-1=69-1=68.

Alternatively, the average of the three consecutive integers is the middle number. Therefore, the middle number = $\frac{207}{3}$ = 69. Hence, the numbers will be 68, 69 and 70. The youngest is 68 years old.

Example 6: A lot of three widgets are stamped with consecutive odd integers. If the numbers on the widgets add up to 93, what is the latest number stamped?

You can assume the numbers as X-2, X and X+2. Sum of the numbers = X-2+X+X+2 = 3X = 93 (Given). Solving for X, X = $\frac{93}{3}$ = 31. The largest number is 31+2=33.

Alternatively, the average of the three numbers is the middle number. Hence, the middle number is $\frac{93}{3}$ = 31. The other two numbers are 29 and 33. The largest number is 33.

Example 7: Two complementary angles are in the ratio of 4:5. What is the larger of the two angles?

Ratio	5	9
Angles in the Triangle	Larger Angle	90

$$\frac{5}{X} = \frac{9}{90}; X = 50.$$

Let the two angles be 4X and 5X. Note that complementary angles add up to 90 degrees. Hence, 4X + 5X = 90; or 9X = 90; solving for X, X = $\frac{90}{9}$ = 10.

Trap: Note that the problem is not asking what X is; the problem is asking what the larger of the two angles is. The larger of the two angles is 5X = 5×10 = 50 degrees

Example 8: My daughter measured angles in a triangle to be in the ratio of 2:3:4. What is the smallest angle measured by my daughter?

Ratio	2	9
Angles in the Triangle	Y	180

$$\frac{2}{Y} = \frac{9}{180}; Y = 40.$$

Let the angles be 2Y, 3Y and 4Y. Angles in a triangle add up to 180 degrees. 2Y + 3Y + 4Y = 180; 9Y = 180; solving for Y, Y = $\frac{180}{9}$ = 20 degrees. The smallest angle of the three is 2Y = 2×20 = 40 degrees. The other two angles are 60 and 80 degrees.

Example 9: Three cities are equidistant from each other. If it takes 36 miles for a round trip to visit all three cities, what is the distance between each of the cities?

Three cities form vertices of an equilateral triangle. Let X be the length of the triangle. In an equilateral triangle, all three sides are of equal length. Hence, X + X + X = 36; 3X = 36; solving for X, X = $\frac{36}{3}$ = 12 miles.

Equations with Two Variables

Equations with two variables have two unknowns in the equation. Note that equations with two variables may not always have a solution. They may also have an infinite number of solutions.

Case of Infinite Solutions (dependent set of equations)

Look at the two equations. X + 3Y = 4 and 3X + 9Y = 12.

Note that 1 and 3 are X coefficients, 3 and 9 are Y coefficients and 4 and 12 are the constants.

	X Coefficient	Y Coefficient	Constant
Equation 1	1	3	4
Equation 2	3	9	12

$$\frac{1}{3} = \frac{3}{9} = \frac{4}{12}$$

SGK's Short Cut: Ratio of coefficients of X and Y = ratio of constants.

Note that the second equation can be derived from the first equation by multiplying the first equation with 3. 3 (X+3Y) = 3×4=12; Hence, the two equations are called dependent equations. So, in reality there is only one equation X+3Y = 4. The equation has infinite solutions because you can pick any value for X and find a corresponding value for Y. (1,1), (4,0), (-5, 3) .. are all solutions of the two equations.

Case of No Solution (contradicting set of equations)

Consider the equations X+3Y = 4 and 3X+9Y = 10.

Note that 1 and 3 are X coefficients, 3 and 9 are Y coefficients and 4 and 10 are the constants.

	X Coefficient	Y Coefficient	Constant
Equation 1	1	3	4
Equation 2	3	9	10

$$\frac{1}{3} = \frac{3}{9} \neq \frac{4}{10}$$

SGK's Short Cut: Ratio of coefficients of X and Y ≠ ratio of constants. Multiplying first equation with 3 yields, 3X+9Y=12 whereas the second equation is 3X+9Y =10; Hence, the two equations contradict each other. No two values of X and Y can satisfy both equations simultaneously.

Case of Unique Solution (determinant of the coefficient matrix is non-zero)

Consider the equations X + 2Y=7 and 3X + Y=6.

Note that 1 and 3 are X coefficients, 2 and 1 are Y coefficients and 7 and 6 are the constants.

	X Coefficient	Y Coefficient	Constant
Equation 1	1	2	7
Equation 2	3	1	6

1×1-2×3=1-6=-5; hence is non-zero. The set of equations has a unique solution. Solving for X and Y, one would get X=1 and Y=3. Note that substituting X=1 and Y=3 satisfies both equations simultaneously.

Solving Equations with Two Variables

One way to solve equations with two variables is by the process of elimination.

In the previous example,

- Multiplying the first equation with 3 yields 3X + 6Y = 21.
- Subtracting second equation from this equation yields 5Y = 15.
- Solving for Y, Y = $\frac{15}{5}$ = 3.
- Substituting Y=3 in any of the two equations yields X=1.

Note the following

- If the problem is asking for a value of X, try to eliminate Y.
- If the problem is asking for a value of Y, try to eliminate X.
- If the problem is not asking for a value of X or Y, maybe there is an easy way to solve the problem. See examples below.
  - Example 10: If X+2Y=5 and 2X+Y=10, what is X+Y?
    - Note that adding the two equations yields 3X+3Y=15.
    - Dividing both sides by 3 yields, X+Y = $\frac{15}{3}$ = 5. Hence, no need to solve for X or Y.
  - Example 11: If X+3Y=4 and 3X+Y=20; what is the average of X and Y?

- Note again, adding the two equations yields 4X+4Y=24; X+Y=6.
  - The average of X and Y is $\frac{(X+Y)}{2} = \frac{6}{2} = 3$.
- If the problem is asking for $\frac{X}{Y}$ or $\frac{Y}{X}$, there is an easier way to solve the problem.
  - Example 12: X+2Y=7 and 3X+Y=6; what is $\frac{X}{Y}$?
    - To solve for $\frac{X}{Y}$, multiply equations with the constant from the other equation and equate the two.
    - 6(X+2Y)=7(3X+Y)
    - 6X+12Y=21X+7Y
    - Rearrange terms to bring all X terms to one side and all Y terms to the other side,
    - 12Y-7Y=21X-6X or 5Y=15X or $\frac{X}{Y} = \frac{5}{15} = \frac{1}{3}$.

Word Problems Involving Two Variables

Example 13: At a 5K run event, there are 120 participants. There are twice as many men as women. How many women participated in the event?

Let the number of men be M and the number of women be W.

M+W=120 and M=2W.

Substituting for M yields, 2W+W=120; 3W=120.

Dividing by 3 yields, $W = \frac{120}{3} = 40$.

Substituting value of W yields, M=2W=2×40=80.

Example 14: A concert charges $3 for children and $6 for adults. If 100 tickets were sold and $450 was collected from the event, how many adult tickets were sold?

Let the number of children be X and the adults be Y.
The two equations are X+Y=100 and 3X+6Y=450.
Multiplying the first equation with 3 yields, 3X+3Y=300.
Subtracting the new equation from the second equation yields, 6Y-3Y=450-300 OR 3Y=150.
Dividing by 3 yields, $Y = \frac{150}{3} = 50$.
Substituting Y=50 in the first equation yields, X+50=100 OR X=100-50=50.

Quadratic Equation

An equation in the form $ax^2 + bx + c = 0$ where $a \neq 0$ is called a quadratic equation.

Properties of a Quadratic Equation

- Solution to a quadratic equation is obtained by using the quadratic formula.

o $X_{1,2} = \frac{-b \pm \sqrt{b^2 - 4ac}}{2a}$.

o The two roots (or solutions) of the equation are $\frac{-b + \sqrt{b^2 - 4ac}}{2a}$ and $\frac{-b - \sqrt{b^2 - 4ac}}{2a}$.

o By adding the two roots, the sum of the roots $= \frac{-b}{2a} + \frac{-b}{2a} = \frac{-b}{a}$.

o By multiplying the two roots, the product of the roots $= \frac{4ac}{4a^2} = \frac{c}{a}$.

- There are three cases.
  o A quadratic equation has two solutions if $b^2 - 4ac > 0$.
  o A quadratic equation has one solution if $b^2 - 4ac = 0$.
  o A quadratic equation has no solution in real X if $b^2 - 4ac < 0$.

- A quadratic equation can be written in the form as $x^2 + \frac{b}{a}x + \frac{c}{a} = 0$. Converting the quadratic equation in this form, one can readily see that
  o The sum of the roots $= \frac{-b}{a}$.
  o The product of the roots $= \frac{c}{a}$.

Example 15: Find out the solutions of the quadratic equation $x^2 - 9x + 20 = 0$.

Using the quadratic formula, a=1, b=-9 and c=20.

Substituting the values in the quadratic formula, $\frac{9 \pm \sqrt{81 - 4 \times 1 \times 20}}{2 \times 1} = \frac{9 \pm \sqrt{81 - 80}}{2} = \frac{9 \pm \sqrt{1}}{2} = \frac{9 \pm 1}{2} = 4$ or 5.

Verify that the sum of the roots = 4 + 5 = 9 = $\frac{-b}{a} = \frac{-(-9)}{1} = 9$.

The product of the roots = 4×5=20= $\frac{c}{a} = \frac{20}{1} = 20$.

Example 16: If p and q are two solutions of the equation $x^2 - 9x + 20 = 0$, what is p+q?

The sum of the roots = p+q= $\frac{-b}{a} = \frac{-(-9)}{1} = 9$.

Example 17: If X=5 is a solution to the equation $X^2 - aX + 20 = 0$, what is another solution to the equation?

The product of the roots = $\frac{c}{a} = \frac{20}{1} = 20$. If X=5 is one solution, another solution should be equal to 4 because 5 × the second solution = 20. The second solution $= \frac{20}{5} = 4$.

Example 18: If X=5 is a solution to the equation $X^2 - 9X + a = 0$, what is another solution to the equation?

The sum of the roots = $\frac{-b}{a} = \frac{-(-9)}{1} = 9$. Hence, the other solution should be 4 because 5 + 4 = 9.

Example 19: If X=4 and X=5 are the solutions to the equation $X^2 - 9X + a = 0$, what is a?

The product of the roots = a = 4×5=20.

Practice Problems

1. How many solutions are possible for the set of equations X+2Y=4 and 3X+6Y=12?
   A. One   B. None   C. Infinite Solutions   D. Cannot be Determined   E. Two

2. How many solutions are possible for the set of equations X+2Y=4 and 3X+6Y=11?
   A. One   B. None   C. Infinite Solutions   D. Cannot be determined   E. Two

3. If X+3Y=6 and 4X+2Y=14; what is X+Y?
   A. 20   B. 5   C. 4   D. 10   E. 8

4. If 5X+2Y=10 and 4X+3Y=5; what is X-Y?
   A. 5   B. 2   C. 3   D. 10   E. 20

5. If X+4=12; what is 3X?
   A. 2   B. 8   C. 24   D. 5   E. 30

6. If $8(\frac{P}{x})+3=27$; what is $\frac{P}{x}$?
   A. 3   B. 4   C. 2   D. 10   E. 5

7. If $8(\frac{P}{x})+3=27$; what is $\frac{x}{P}$?
   A. 4   B. 3   C. $\frac{1}{3}$ D. $\frac{1}{4}$ E. $\frac{1}{5}$

8. If X+2Y=5 and 3X+4Y=11; what is $\frac{X}{Y}$?
   A. 1   B. 2   C. $\frac{1}{2}$ D. $\frac{1}{3}$ E. $\frac{1}{4}$

9. If X+5=12; what is 4X+20?
   A. 20   B. 30   C. 36   D. 48   E. 50

10. If 3X+15=27; what is X+5?
    A. 15   B. 20   C. 12   D. 10   E. 9

11. If 3X+15=27; what is X?
    A. 3   B. 6   C. 4   D. 5   E. 9

12. An amusement park is issuing tokens consisting of consecutive odd integers to its visitors. The sum of last three tokens issued is 99. What is the last token issued?
    A. 31   B. 33   C. 35   D. 37   E. 32

13. The sum of 11 test scores equals 99. What is the average score?
    A. 33   B. 22   C. 9   D. 11   E. 10

14. The sum of 9 numbers is 0. What is the average?
    A. 9   B. 7   C. 5   D. 3   E. 0

15. A machine is generating consecutive odd integers. The sum of last 8 numbers generated is 0. What is the smallest number out of these 8 numbers?
    A. -5   B. 5   C. -7   D. 7   E. 0

16. If 2X+3Y=7 and 3X+2Y=3, what is X?

    A. 1   B. -1   C. 0   D. 3   E. -3

17. If X+4Y = 8 and 2X+3Y=6, what is X?

    A. 1   B. 3   C. 0   D. 4   E. -1

18. Jack is thrice as old as Jill. In two years, Jack will be twice as old as Jill. How old will be Jack in five years?

    A. 6   B. 8   C. 11   D. 12   E. 4

19. If $m + \frac{1}{m} = 7$, what is $m^2 + \frac{1}{m^2}$?

    A. 9   B. 49   C. 47   D. 51   E. 81

20. If X=4 is a solution to the equation $X^2 - aX + 8 = 0$, what is another solution to the equation?

    A. 2   B. -2   C. 0   D. 4   E. -4

21. If X=4 is a solution to the equation $X^2 - aX + 8 = 0$, what is a?

    A. 2   B. -2   C. 6   D. 4   E. -4

22. If p and q are solutions to the equation $X^2 - aX + 8 = 0$, what is pq?

    A. 2   B. -2   C. 6   D. 8   E. -8

23. If p and q are solutions to the equation $X^2 - 6X + a = 0$, what is p+q?

    A. 2   B. -2   C. 6   D. 8   E. -8

Solutions to Practice Problems

1. **Best answer is C.** The coefficient matrix is $\begin{vmatrix} 1 & 2 \\ 3 & 6 \end{vmatrix}$. The determinant = 1×6-2×3=6-6=0. The ratio of coefficients of X and Y= the ratio of constants. Also, note that multiplying the 1st equation with 3 yields 3x+6y=12; same as the second equation. Hence, we have only one equation, X+2Y=4. One can choose many values for X and Y. For example, (0,2), (2,1), 4,0)... Hence, there are infinite solutions.

    SGK's Short Cut: The coefficients of X and Y are in proportion to the ratio of constants. i.e., $\frac{1}{3} = \frac{2}{6} = \frac{4}{12}$. Hence, there are infinite solutions.

	X Coefficient	Y Coefficient	Constant
Equation 1	1	2	4
Equation 2	3	6	12

2. **Best answer is B.** The coefficient matrix is $\begin{vmatrix} 1 & 2 \\ 3 & 6 \end{vmatrix}$. The determinant = 1×6-2×3=6-6=0. However, the ratio of coefficients of X and Y is not equal to the ratio of constants. Also, note that multiplying the 1st equation with 3 yields 3X+6Y=12; whereas the second equation is 3X+6Y=11. Hence, the two equations contradict each other. No solution can satisfy both the equations <u>simultaneously</u>.

    SGK's Short Cut: The coefficients of X and Y are not in proportion to the ratio of constants. i.e., $\frac{1}{3} = \frac{2}{6} \neq \frac{4}{11}$. Hence, there are not solutions.

	X Coefficient	Y Coefficient	Constant
Equation 1	1	3	4
Equation 2	3	9	11

3. **Best answer is C.** *SGK's Short Cut:* No need to solve for X and Y. Adding both equations yields 5X+5Y=20; or 5(X+Y)=20 and X+Y=$\frac{20}{5}$ = 4.

4. **Best answer is A.** *SGK's Short Cut:* Subtracting the second equation from the first equation yields X-Y=5. Once again, no need to solve for X and Y.

5. ***Best answer is C.*** *Trap*: problem is asking for 3X, not X. X=12-4=8, hence 3X=3×8=24. Note that the problem is asking for 3X. Hence, choices A, B and D can be eliminated because they are not multiples of 3. Only choices C and E are multiples of 3.

6. ***Best answer is A.*** No need to panic. Treat $\left(\frac{p}{x}\right)$ as a variable. Let us say, $\frac{p}{x}$=Y, the equation is 8Y+3=27; 8Y=27-3=24, solving for Y, Y = $\frac{24}{8}$ = 3. Because, Y is same as $\frac{p}{x}$, $\frac{p}{x}$=3.

7. ***Best answer is C.*** *Trap:* The question is asking for $\frac{x}{p}$, and not $\frac{p}{x}$. Solve for $\frac{p}{x} = \frac{27-3}{8}$ =3. Take the inverse. If $\frac{p}{x}$=3, $\frac{x}{p}$ is $\frac{1}{3}$.

8. ***Best answer is C.*** *SGK's Short Cut:* Using the technique illustrated in the example 12, 11(X+2Y)=5(3X+4Y); 11X+22Y=15X+20Y; re arranging terms, 15X-11X=22Y-20Y; 4X=2Y; $\frac{X}{Y} = \frac{2}{4} = \frac{1}{2}$.

9. ***Best answer is D.*** Note that 4X+20=4(X+5); hence, 4X+20=4×12=48. No need to solve for X.

10. ***Best answer is E.*** Note that 3X+15=3(X+5). Hence, X+5=$\frac{27}{3}$ = 9.

11. ***Best answer is C.*** If 3X+15=27, X+5=9; hence X=9-5=4.

12. ***Best answer is C.*** The middle number = the average of the numbers = $\frac{99}{3}$ = 33. Hence, the numbers are 31, 33, and 35.

13. ***Best answer is C.*** The average = $\frac{\text{The sum of the numbers}}{11} = \frac{99}{11} = 9$.

14. ***Best answer is E.*** The average = $\frac{\text{The sum of the numbers}}{9} = \frac{0}{9} = 0$.

15. ***Best answer is C.*** Because there are 8 odd integers, there is no middle number. However, the two middle numbers average to 0. Hence, the numbers in the middle should be -1 and 1. The numbers in the series are -7, -5, -3, -1, 1, 3, 5, and 7. The smallest number is -7.

16. ***Best answer is B.*** Note that the coefficients of X and Y are interchanged in both the equations. Use the following technique. Step 1: Add both equations to get 5X+5Y=10 or X+Y=2. Step 2: Subtract second equation from the first equation to get −X+Y=4. Now, solving for X and Y, X=-1 and Y=3.

17. ***Best answer is C.*** Note that the Y coefficient and the constant term are in proportion. $\frac{4}{8} = \frac{3}{6}$. If one of the coefficients and the constant term are in proportion, the other variable is zero. Hence, X=0. (and Y=2).

18. ***Best answer is C.*** Let Jack is X years old and Jill is Y years old. X=3Y and (X+2)=2(Y+2). Solving for X and Y, X=2 and Y=6. *Trap:* The problem is asking for Y+5=6+5=11.

19. ***Best answer is C.*** $\left(m + \frac{1}{m}\right)^2 = m^2 + 2 + \frac{1}{m^2} = 7^2 = 49$. Hence, $m^2 + \frac{1}{m^2} = 49 - 2 = 47$.

20. ***Best answer is A.*** The product of the roots $= \frac{c}{a} = \frac{8}{1} = 8$. Because X=4 is one solution, the other solutions should be 2. 2×4=8.

21. **Best answer is C.** Based on the previous problem, the other solution is X=2. The sum of the roots = 2+4=6. Hence, a=6.

22. **Best answer is D.** The product of the roots = pq = 8.

23. **Best answer is C.** The sum of the roots = p+q = 6.

7. Exponents

Exponents

Let us study the number b^x. "b" is called the "base". "x" is called the "exponent". To simplify exponent problems, you need to know the following rules.

Rule 1 – Same Base

Multiplication: When <u>bases</u> are same, add the <u>exponents</u>. See examples below.

$$b^x \times b^y = b^{x+y} \qquad\qquad 2^3 \times 2^4 = 2^{3+4} = 2^7 \qquad 3^2 \times 3^9 = 3^{2+9} = 3^{11}$$

Rule 2 – Same Exponents

Multiplication: When <u>exponents</u> are same, multiply the <u>bases</u>. See examples below.

$$b^x \times a^x = (b \times a)^x \qquad\qquad 2^4 \times 3^4 = (2 \times 3)^4 = 6^4$$

Rule 3 – Negative Exponents

Division: Use the reciprocal to convert a negative exponent into positive exponent (and vice versa).

$$b^{-x} = \frac{1}{b^x} \qquad\qquad\qquad \frac{1}{b^{-x}} = b^x$$

$$2^{-3} = \frac{1}{2^3} = \frac{1}{8} \qquad\qquad \frac{2^4}{2^3} = 2^{4-3} = 2^1 = 2$$

Rule 4 – Exponent of Exponents

$$(b^x)^y = b^{xy} \qquad\qquad\qquad (2^3)^4 = 2^{3 \times 4} = 2^{12}$$

Rule 5 – Roots

Square root is same as exponent $\frac{1}{2}$. Cube root is same as exponent $\frac{1}{3}$. Nth root is same as exponent $\frac{1}{n}$.

$$\sqrt[3]{27} = 27^{\left(\frac{1}{3}\right)} = (3^3)^{\left(\frac{1}{3}\right)} = 3^{\left(3 \times \frac{1}{3}\right)} = 3^1 = 3.$$

Rule 6 – Exponent=0

Any number to the power of zero is equal to 1. $2^0 = 3^0 = x^0 = 1$.

<u>Example 1:</u> What is $(3x^3)^5$?

It is simplified as $3^5(x^3)^5 = 243(x^{15})$.

<u>Example 2:</u> What is $\sqrt[3]{27x^3y^9z^4}$?

27 is 3^3. The expression simplifies to $3xy^3z\sqrt[3]{z}$.

Example 3: Simplify $\dfrac{x^2 y^4 z^7}{x^3 y^8 z^3}$.

The expression simplifies to $x^{(2-3)} y^{(4-8)} z^{(7-3)} = x^{-1} y^{-4} z^4 = \dfrac{z^4}{xy^4}$.

Example 4: Simplify $\dfrac{x^4 y^9 z - x^3 y^2 z^9}{xy}$.

The expression simplifies to $\dfrac{x^4 y^9 z}{xy} - \dfrac{x^3 y^2 z^9}{xy} = x^3 y^8 z - x^2 yz^9$.

Example 5: Solve for x, if $25^{(2x-5)} = 125^{3x}$.

Bring both sides to a common base. In this case, common base is 5. You find out the common base by using prime factorization. $25 = 5 \times 5 = 5^2$; whereas $125 = 5 \times 5 \times 5 = 5^3$.
$25^{(2x-5)} = (5^2)^{(2x-5)} = 5^{2(2x-5)} = 5^{4x-10}$.
$125^{3x} = (5^3)^{3x} = 5^{3 \times 3x} = 5^{9x}$.
Hence, $5^{4x-10} = 5^{9x}$.

Because the bases are the same, equating the exponents yields 4x-10 = 9x; solving for x, x = -2.

Practice Problems

1. What is $(3x^2)^4$?
   A. $3x^8$ B. $27x^4$ C. $81x^8$ D. $3x^8$ E. $81x^2$

2. What is $(-2x^2)^5$?
   A. $-2x^{10}$ B. $2x^{10}$ C. $-32x^2$ D. $32x^{10}$ E. $-32x^{10}$

3. Simplify $(3x^7)\times(4x^4)$.
   A. $12x^{11}$ B. $12x^{28}$ C. $7x^{28}$ D. $7x^{12}$ E. $12x^{10}$

4. What is $(-5x^{-2})\times(6x^3)$?
   A. $30x^5$ B. $30x$ C. $-30x$ D. $\frac{-30}{x}$ E. $\frac{30}{x}$

5. Simplify $2\times(3x^2)^3$.
   A. $6x^6$ B. $54x^6$ C. $54x^9$ D. $18x^6$ E. $18x^9$

6. If $x = 3$ and $y = 2$, what is $(x + y)^3$?
   A. 5 B. 25 C. 125 D. 625 E. 10

7. If $x = 2$ and $y = 3$, what is x^3y^2?
   A. 72 B. 24 C. 108 D. 54 E. 6

8. Simplify $\frac{2x^3\times18x^2}{4x^7}$.
   A. $9x^2$ B. $\frac{8}{x^2}$ C. $\frac{9}{x^2}$ D. $9x^3$ E. $36x$

9. Simplify $\frac{4x^5-2x^7}{2x^3}$.
   A. $2x^2 - x^4$ B. $2x^2 - 2x^4$ C. x^2 D. x^4 E. $\frac{2x^3-x}{x^2}$

10. What is $\frac{20x^2y^7z^3}{25x^3y^3z^9}$?
    A. $\frac{5xy^2z^3}{4}$ B. $\frac{4xy^2z^3}{5}$ C. $\frac{5y^4}{4xz^6}$ D. $\frac{4y^4}{5xz^6}$ E. $\frac{5}{4xyz}$

11. If $4^{2x+3} = 64^x$, what is x?
    A. 1 B. 3 C. 2 D. 4 E. 0

12. What is $625^{\frac{3}{4}}$?
    A. 5 B. 25 C. 125 D. 625 E. 2025

13. What is $x^{\frac{1}{2}}x^{\frac{1}{3}}$?
    A. $x^{\frac{1}{6}}$ B. x C. $x^{\frac{5}{6}}$ D. $x^{\frac{3}{5}}$ E. x^2

14. If $x = 2$, $y = 3$ and $z = 5$, what is $(x+y-z)(x-y+z)$?
    A. 30 B. 40 C. 20 D. 0 E. 10

15. If $x=3$; $y=2$, $z=7$; what is $(x+y+z)(x-y+z)$?
    A. 101 B. 39 C. 96 D. 95 E. 121

16. If x=3, y=2 and z=5; what is $(x+y+2z)(x+y-2z)$?

    A. -75   B. 75   C. 25   D. 125   E. 55

17. If x=4, y = 6 and z=5; what is $(x+y+z)(x-y+z)$?

    A. 40   B. 50   C. 60   D. 45   E. 80

Solutions to Practice Problems

1. **Best answer is C.** First look at the constant term. $3^4 = 3 \times 3 \times 3 \times 3 = 81$. Hence, you can eliminate all choices except C and E. $(x^2)^4 = x^{2 \times 4} = x^8$. Hence, Chioce C is the best answer.

2. **Best answer is E.** Similar to the previous problem, note that $(-2)^5 = (-2) \times (-2) \times (-2) \times (-2) \times (-2) = -32$. $(x^2)^5 = x^{2 \times 5} = x^{10}$.

3. **Best answer is A.** The constant term is 3×4=12. $x^7 x^4 = x^{7+4} = x^{11}$. Because the base is same, add the exponents.

4. **Best answer is C.** The constant term is -5×6=-30. Eliminate all choices except C and D. Add the exponents -2+3=1. C is the best choice.

5. **Best answer is B.** The constant term is 2×3×3×3=54. Eliminate all choices except B and C. $(x^2)^3 = x^{2 \times 3} = x^6$.

6. **Best answer is C.** Substituting values for x and y, x+y=3+2=5. $(x + y)^3 = 5^3 = 5 \times 5 \times 5 = 125$.

7. **Best answer is A.** Substituting values for x and y, $x^3 y^2 = 2^3 \times 3^2 = 8 \times 9 = 72$.

8. **Best answer is C.** The constant term is $\frac{2 \times 18}{4} = 9$. Note that $\frac{1}{x^7} = x^{-7}$. Adding exponents, 3+2-7=-2. $x^{-2} = \frac{1}{x^2}$.

9. **Best answer is A.** $\frac{4x^5 - 2x^7}{2x^3} = \frac{4x^5}{2x^3} - \frac{2x^7}{2x^3} = 2x^{5-3} - x^{7-3} = 2x^2 - x^4$.

10. **Best answer is D.** The constant term is $\frac{20}{25} = \frac{4}{5}$; hence, eliminate all choices except B and D. Choice B is wrong because all exponents are in the numerator.

11. **Best answer is B.** Note that 4 and 64 are both divisible by 2; hence 2 is a common base. Convert the equation into base 2. $4^{2x+3} = (2^2)^{2x+3} = 2^{2(2x+3)} = 2^{4x+6}$. $64^x = (2^6)^x = 2^{6x}$. Comparing the exponents, 4x+6=6x. Solving for x, 6x-4x=6 and 2x=6. $x = \frac{6}{2} = 3$.

12. **Best answer is C.** $625^{\frac{3}{4}} = (5^4)^{\frac{3}{4}} = 5^{(4 \times \frac{3}{4})} = 5^3 = 125$.

13. **Best answer is C.** Adding the exponents, $\frac{1}{2} + \frac{1}{3} = \frac{3+2}{6} = \frac{5}{6}$.

14. **Best answer is D.** *SGK's Short Cut:* It is a trick question. Note that x+y-z=2+3-5=0. It does not matter what x-y+z is, the answer is going to be 0 because 0 multiplied by any number is 0.

15. **Best answer is C.** *SGK's Short Cut:* It is a trick question. Note that x+y-z=3+2+7=12. None of the choices except Choice C is a multiple of 12.

16. **Best answer is A.** x+y+2z=3+2+10=15 and x+y-2z=3+2-10=-5. Hence, the answer is 15x-5=-75. Note that there is only one negative number in the answers. You can pick the choice without multiplying 15 and -5.

17. **Best answer is D.** x+y+z=4+6+5=15 and x-y+z=4-6+5=3. Hence, the answer is 15×3=45.

8. Inequalities

Inequalities involve four types of problems. They are ">" greater than, "<" less than, ">=" greater than or equal to, "<=" less than or equal to. ">" and "<" does not include the end point where as ">=" and "<=" include the end point. Solving inequalities involve the same rules as solving equations. However, there is one very important rule to remember. When an inequality is multiplied or divided by a negative number, the inequality changes.

For example, x < 2 means –x > -2. To solve, -2x < 6, one would divide the inequality by -2. Hence, one would get x > -3. Notice the sign has changed from "<" to ">".

Example 1: If $2X - 6 < 9$, solve for X.

$$2X - 6 + 6 < 9 + 6; \ 2X < 15; X < \frac{15}{2}; X < 7.5$$

This can be represented in the range as (-∞, 7.5). Use open parenthesis "( )" when the end points are not included. Also, open parenthesis is used when "infinite" is included. Closed parenthesis is used when the end points are included. Closed parenthesis is <u>never</u> used when "infinite" is included. See the table below.

X<2	(-∞, 2)
X≤2	(-∞, 2]
X>2	(2,∞)
X≥2	[2,∞)
-1<X<3	(-1,3)
-1≤X<3	[-1,3)
-1<X≤3	(1,3]
-1≤X≤3	[-1,3]

Example 2: If $8 - 3X > -7$, solve for X.

$8 - 3X > -7; \ 8 - 3X - 8 > -7 - 8; \ -3X > -15; 3X < 15; X < 5$. i.e., X is (-∞,5).

Example 3: If $2X - 5 \geq 3X - 9$, solve for X.

$2X - 5 \geq 3X - 9; 2X - 5 + 9 \geq 3X - 9 + 9; 2X + 4 \geq 3X; 2X + 4 - 2X \geq 3X - 2X; 4 \geq X$ or $X \leq 4$.

Example 4: If $2^X = 100$, what is X?

Note that $2^2 = 4; \ 2^3 = 8; 2^4 = 16, 2^5 = 32; \ 2^6 = 64$ and $2^7 = 128$. Because 100 is between 64 and 128, X must be between 6 and 7. Hence, 6<X<7.

Example 5: Solve for X given $3X - 7 < 2X + 8 < 5X - 4$.

Separate these into two inequalities and then, solve. Solving the first inequality, $3X - 7 < 2X + 8; X < 15$. Solving the second inequality, $2X + 8 < 5X - 4; 8 + 4 < 5X - 2X; 12 < 3X; 4 < X$ or $X > 4$.

Now, combining both solutions results in $4 < X < 15$.

Practice Problems

1. If $2X - 12 > -6$, solve for X.

    A. X < 9   B. X > 9   C. X < 3   D. X > 3   E. X=3

2. If $\frac{2X-3}{3} < 5$, solve for X.

    A. X < 10   B. X > 9   C. X < 9   D. X> 10   E. X<5

3. $3X - 5 \geq -9$, solve for X.

    A. $X \geq \frac{-4}{3}$ B. $X \leq \frac{-4}{3}$ C. $X \geq \frac{4}{3}$ D. $X \leq \frac{4}{3}$ E. $X \leq \frac{14}{3}$

4. $X - 2 < 2X + 3 < X + 11$, solve for X.

    A. -10<X<5   B. -5 < X < 8   C. -5>X>8   D. -8<X<5   E. -8>X>5

5. What is one value of X that satisfies $4X - 5 \leq 7$.

    A. 4   B. 5   C. 3   D. 10   E. 12

6. If Sam has three dollars and each pencil costs 35 cents, how many pencils can Sam buy?

    A. 10   B. 20   C. 8   D. 9   E. 10

7. A 6 inch diameter bolt is to be made with a tolerance of 0.005 inch. What is the acceptable range for the diameter of the bolt?

    A. 5.995 to 6.5   B. 6 to 6.005   C. 5.995 to 6   D. 5.995 to 6.005   E. 5.95 to 6.05

8. A 5 inch diameter bolt is to be made with a tolerance of 0.01 inch. Which of the following bolts shall be rejected? One with a diameter of:

    A. 5.005   B. 4.995   C. 4.97   D. 4.997   E. 5

9. A 3 inch diameter bolt is to be made with a tolerance of 0.002 inch. Which of the following bolts shall be accepted? One with a diameter of:

    A. 2.995   B. 3.003   C. 3.01   D. 3.0005   E. 2.99

Solutions to Practice Problems

1. ***Best answer is D***. $2X - 12 > -6$; $2X - 12 + 12 > -6 + 12$; $2X > 6$; $X > 3$.

2. ***Best answer is C***. $\frac{2X-3}{3} < 5$; $2X - 3 < 15$; $2x - 3 + 3 < 15 + 3$; $2X < 18$; $X < 9$.

3. ***Best answer is A***. $3X - 5 \geq -9$; $3X - 5 + 5 \geq -9 + 5$; $3X \geq -4$; $X \geq \frac{-4}{3}$.

4. ***Best answer is B***. Solve one at a time.

    $X - 2 < 2X + 3$; $-5 < X$ or $X > -5$;

    $2X + 3 < X + 11$; $X < 8$;

    Combining both the equations, $-5 < X < 8$.

5. ***Best answer is C***. $4X - 5 \leq 7$; $4X - 5 + 5 \leq 7 + 5$; $4X \leq 12$; $X \leq 3$. Because X can be less than or equal to 3, X=3 is a valid value.

6. ***Best answer is C***. $X \leq \frac{300}{35}$; $X \leq 8.57$. You cannot buy half a pencil, hence X = 8.

7. ***Best answer is D***. $6 - 0.005 \leq X \leq 6 + 0.005$; $5.995 \leq X \leq 6.005$.

8. ***Best answer is C***. $5 - 0.01 \leq X \leq 5 + 0.01$; $4.99 \leq X \leq 5.01$. 4.97 is out of the range and hence, it shall be rejected.

9. ***Best answer is D***. $3 - 0.002 \leq X \leq 3 + 0.002$; $2.998 \leq X \leq 3.002$. Choice D 3.0005 is within the range.

9. Absolute Value

Absolute value refers to the magnitude of a number. It represents the distance of the number from 0. Absolute value is <u>always positive</u>. For example, absolute value of 5 is 5. On the other hand, absolute value of -5 is 5. Equations and inequalities involving absolute values are studied more easily if one can think of them as circle problems. For example, $|x| < 4$ represents a circle with center at x=0 and radius=4 units. Solution for the inequality is -4 < x < 4. i.e., all points along the x-axis that are between -4 and 4. Similarly, $|x| > 4$ means x < -4 or x > 4. i.e., points outside the circle.

<u>Example 1:</u> $|x - 3| < 5$

Method 1: Separate this into two inequalities $X - 3 < 5$ or $X - 3 > -5; X < 8$ or $X > -2; -2 < X < 8$

SGK's Short Cut: Think of this as a circle problem with center at X-3=0 or X=3 and radius = 5.

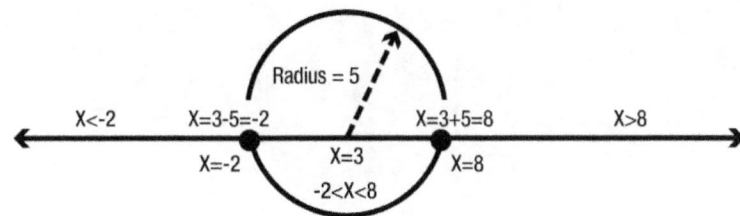

Note that X=3-5 and X=3+5 are two end points of the diameter along the X-axis. Therefore, X=-2 and X=8. Hence, X should be between -2 and 8.

<u>General Steps</u>

<u>Step 1:</u> Convert the problem into the form $|X - c|$? r

<u>Step 2:</u> Solve for X-c=0 i.e., X=c.

<u>Step 3:</u> Note that "r" is the radius.

<u>Step 4:</u> Find the boundaries by calculating X=c+r and X=c-r.

<u>Step 5:</u> If the problem involves "<", then X is within (c-r,c+r). If the problem involves ">", then X is outside (c-r,c+r).

<u>Example 2:</u> $3|X - 2| > 6$

Divide both sides of the equation by 3. Simplify this as $|X - 2| > \frac{6}{3}$; $|X - 2| > 2$. Solving X-2=0; X =2 is the center of the circle. And the radius of the circle is 2. Note c=2 and r=2. 2-2 = 0 and 2+2 = 4. Hence, X < 0 or X > 4 is the solution for the inequality.

<u>Example 3:</u> $|3X - 6| > 8$

Convert the equation to $3|X - 2| > 8$; $|X - 2| > \frac{8}{3}$.

Solve X-2 = 0; X=2. X=2 represents the center of the circle. Radius = $\frac{8}{3}$. c=2 and r=$\frac{8}{3}$. Hence, X< $2 - \frac{8}{3}$ or X> $2 + \frac{8}{3}$. i.e., X< $-\frac{2}{3}$ or X > $\frac{14}{3}$.

Example 4: $|2X - 3| < 4$

Convert the equation to $2\left|X - \frac{3}{2}\right| < 4$; $\left|X - \frac{3}{2}\right| < 2$.

Solve $X - \frac{3}{2} = 0$; $X = \frac{3}{2} = 1.5$.

X = 1.5 is the center of the circle and radius=2. i.e., c=1.5 and r=2. So, 1.5-2 and 1.5+2 are the two ends of the diameter. X=-0.5 and X=3.5 are the two ends of the circle. Therefore, -0.5 < X < 3.5 is the solution of the inequality.

Practice Problems

1. If $|x - 3| = 2$, what is x?

   A. x=5   B. x=±5   C. x=5 or x=1   D. x=3   E. x=2 or x=3

2. If $|2x - 5| < 3.5$, what is x?

   A. x>0.75 or x <4.25   B. x < 0.75 or x > 4.254

   C. x < 0.75 and x >4.25   D. x>0.75 and x<4.25   E. x=0.75 or x=4.25

3. If $|x - 7| > 2$, what is x?

   A. x < 5 or x >9   B. 5 < x < 9   C. x=5 or x =9   D. x=2 or x=7   E. 2<x<7

4. If $\left|\dfrac{x}{2} - 4\right| < 3$, what is x?

   A. 2<x<14   B. x=2 or x=14   C. x<2 or x>14   D. x>7   E. 5<x<11

5. If $|x + 4| - 3 < 5$, what is x?

   A. x=4 or x=12   B. x<4 or x>12   C. 4<x<12   D. -4<x<-12   E. -12<x<4

68

Solutions to Practice Problems

1. ***Best answer is C.*** x-3=2 or x-3=-2; x=5 or x=1. Or think of this as a circle with the center at x=3 and radius 2 units. So, the solution is 3+2=5 or 3-2=1.

2. ***Best answer is D.*** Convert the equation to $2|x - 2.5| < 3.5$; $|x - 2.5| < 1.75$. Solving, x-2.5=0, x=2.5. i.e., c=2.5 and r=1.75. The center of the circle is at x=2.5 and radius is 1.75. Hence, the two end points are 2.5+1.75=4.25 and 2.5-1.75=0.75. Hence, 0.75<x<4.25. Note that Choices A and D look similar. Choice A is wrong. "OR" means x>0.75 which means, x can be greater than 4.25. The statement should say "AND" to limit the values to be between 0.75 and 4.25. Choice C is also wrong because x cannot be less than 0.75 and greater than 4.25 at the same time.

3. ***Best answer is A.*** x-7=0 means x=7 is the center of the circle. Radius=2 units. So, the end points are 7+2=9 and7-2=5 units. Hence, the solution is x<5 or x>9. Unlike problem 2, in this case "OR" is appropriate, not "AND".

4. ***Best answer is A.*** Convert the equation to $2\left|\frac{x}{2} - 4\right| < 6$; $|x - 8| < 6$. x=8 is the center of the circle and radius=6 units. i.e., c=8 and r=6. Hence, the two end points are 8+6=14 and 8-6=2. The solution is 2<x<14.

5. ***Best answer is E.*** $|x + 4| - 3 < 5$; $|x + 4| < 5 + 3$; $|x + 4| < 8$. x=-4 is the center of the circle and radius is 8 units. i.e., c=-4 and r=8. Hence, the two end points are -4+8=4 and -4-8=-12. The solution is -12<x<4.

10. Arithmetic and Geometric Series

Arithmetic Series

How can you identify an arithmetic series? An arithmetic series is one where each term differs the previous term by a constant value.

Example 1: 2, 5, 8, 11, 14, 17... Each term differs the previous term by 3.

Example 2: 1, -4, -9, -14, -19... Each term differs the previous term by -5.

Properties of an Arithmetic Series

Arithmetic series have a very important characteristic. *Each term equals the average of the two adjacent terms.* In the example 1 above, note the following:

2nd term = $\frac{2+8}{2}$ = 5; 3rd term = $\frac{5+11}{2}$ = 8; 4th term = $\frac{8+14}{2}$ = 11.

General Arithmetic Series

Consider the arithmetic series $a_1, a_2, a_3, ..., a_{n-1}, a_n$.

Initial Term or First Term

An arithmetic series starts with a number. The first number in the series is called the "Initial Term" or the "First Term". "2" is the first term in the example 1 and "1" is the first term in the example 2.

Common Difference

The common difference is the difference between the current term and the previous term. When determining the common difference, direction is important.

The common difference = $a_2 - a_1 = a_3 - a_2 = a_4 - a_3 = \cdots = a_n - a_{n-1}$

Finding the nth term in an Arithmetic Series

Let us say the first term is "a" and the common difference is "d".

The second term = a+d; The third term = a+d+d = a+2d; The fourth term = a+2d+d=a+3d

Similarly, the nth term = a+(n-1)d

In the example 1, the nth term = $2 + (n-1)3 = 2 + 3n - 3 = 3n - 1$.

In the example 2, the nth term = $1 + (n-1) \times (-5) = 1 - 5n + 5 = 6 - 5n$.

Finding the Sum of n Terms in an Arithmetic Series Given the First and Last Terms

Let us say a_1 is the first term and a_n is the last term. One can use the property of an arithmetic series to find the sum of the first n terms. The average of the first and last terms $= \frac{a_1 + a_n}{2}$

Because there are n terms in the series, the sum of the first n terms $= n \times (\frac{a_1 + a_n}{2})$.

Finding the Sum of n Terms in an Arithmetic Series Given the First Term and the Common Difference

Find the nth term $= a_1 + (n-1)d$.

The sum of the first n terms $= n\left(\frac{a_1 + a_n}{2}\right) = \frac{n}{2}(a_1 + a_1 + (n-1)d) = \frac{n}{2}(2a_1 + (n-1)d)$.

Example 3: Find the sum of first 200 integers.

The first term = 1; the nth term = 200 and n = 200.

Hence, the sum of first 200 integers $= \frac{200}{2}(1 + 200) = 100 \times 201 = 20{,}100$.

Example 4: Find of the sum of first 100 odd integers.

Odd integers are 1, 3, 5, ... There are 100 odd integers; hence, n = 100. The first term = 1 and the last term is 199. The sum of first 100 odd integers $= \frac{100}{2}(1 + 199) = 50 \times 200 = 10{,}000$.

Example 5: Find the sum of first 20 terms of the series 1, 4, 7, 10, 13...

The first term is 1, the common difference is 4-1=3; n = 20. The sum of first 20 terms $= \frac{20}{2}(2 \times 1 + (19 \times 3)) = 10 \times (2 + 57) = 590$.

Example 6: How many terms are in the series 2, 8, 14, ..., 68?

The first term, a=2, The common difference, d=8-2=6; the nth term, $a_n = 68 = 2 + (n-1)6$;

$n - 1 = \frac{68-2}{6} = \frac{66}{6} = 11; n = 11 + 1 = 12$. Hence, there are 12 terms in the series.

Example 7: Find the sum of 3, 7, 11, 15, ..., 83.

First find out how many terms are in the series. The nth term $= 83 = 3 + (n-1)4$; solving for n, n=21. The sum of first 21st terms $= \frac{21}{2}(3 + 83) = \frac{21}{2} \times 86 = 21 \times 43 = 903$.

Geometric Series

How can you identify a geometric series? A geometric series is one where the ratio of two successive terms is constant. The ratio of two successive terms in a geometric series is called the common ratio.

Example 8: 2, 6, 18, 54, ... The common ratio $= \frac{6}{2} = \frac{18}{6} = \frac{54}{18} = 3$.

Example 9: 1, -2, 4, -8, 16... The common ratio = $\frac{-2}{1} = \frac{4}{-2} = \frac{-8}{4} = -2$.

Properties of a Geometric Series

Geometric series have a very important characteristic. <u>In a geometric series, each term is the geometric average of the two adjacent terms.</u> The geometric mean of two numbers is calculated by calculating the square root of the product of the two numbers. In the example 8 above, note the following:

The 2nd term = $\sqrt{2 \times 18} = \sqrt{36} = 6$.

The 3rd term = 18 = $\sqrt{6 \times 54} = \sqrt{324} = 18$.

General Geometric Series

Initial Term or First Term

Geometric series starts with a number. The first number in the series is called the "Initial Term" or the "First Term". "2" is the first term in the example 8 and "1" is the first term in the example 9.

Common Ratio

The Common Ratio is the ratio between the current term and the previous term. When determining the common ratio, direction is important.

The common ratio = $\frac{a_2}{a_1} = \frac{a_3}{a_2} = \frac{a_4}{a_3} = \cdots = \frac{a_n}{a_{n-1}}$.

Finding the nth Term in a Geometric Series

Let us say the first term is "a" and the common ratio is "r".

The 2nd term is a×r = ar; the 3rd term is a×r×r = ar^2; the 4th term is a×r×r×r = ar^3.

In general, the nth term of a geometric series = ar^{n-1}.

In the example 8, the nth term = $2 \times 3^{n-1}$.

In the example 9, the nth term = $1(-2)^{n-1} = (-2)^{n-1}$.

Identifying An Arithmetic or A Geometric Series

Sometimes, the problem does not state whether the given series is an arithmetic series or a geometric series. Here is how you will find out.

1. Test for arithmetic series by
   a. calculating $a_2 - a_1$, $a_3 - a_2$, $a_4 - a_3$. If they are all equal, it is an arithmetic series. In addition, note that in an arithmetic series, each term is an arithmetic average of the two adjacent terms.

2. Test for geometric series by
   a. calculating $\frac{a_2}{a_1}, \frac{a_3}{a_2}, \frac{a_4}{a_3}$. If they are all equal, it is a geometric series. In addition, note that in a geometric series, each term is a geometric average of the two adjacent terms.

<u>Example 10:</u> Identify the series. 11, 14, 17, 20, 23…

14-11=17-14=20-17=23-20=3; hence it is an arithmetic series. In addition, note that $\frac{11+17}{2} =$ 14; $\frac{14+20}{2} = 17$; $\frac{17+23}{2} = 20$.

<u>Example 11:</u> Identify the series, 2, $\frac{-1}{3}, \frac{1}{18}, \frac{-1}{108}$, …

$\frac{\frac{-1}{3}}{2} = \frac{-1}{6} = \frac{\frac{1}{18}}{\frac{-1}{3}} = \frac{-1}{6}$; hence, it is a geometric series.

SGK's Short Cut: Note also that an <u>alternating series (alternating positive and negative terms) is likely to be a geometric series.</u>

Finding the sum of n Terms in a Geometric Series Given the First Term and the Common Ratio

Let us say "a" is the first term and "r" is the common ratio.

The sum of first "n" terms of a geometric series = a.$\frac{r^n-1}{r-1}$.

<u>Example 12:</u> What is the sum of geometric series 2, $\frac{2}{3}, \frac{2}{9}, \frac{2}{27}$ … up to 10 terms?

The first term a = 2, the common ratio r $= \frac{1}{3}$, n = 10. The sum of first 10 terms = $2 \times \frac{\frac{1}{3}^{10}-1}{\frac{1}{3}-1}$=3.

<u>Special case:</u> When "r" is between -1 and 1 i.e., -1 < r < 1 or $|r| < 1$, Sum of geometric series for infinite terms (all the terms) = $\frac{a}{1-r}$. Note that r could be negative.

<u>Example 13:</u> What is the sum of the series 1, $\frac{1}{2}, \frac{1}{4}, \frac{1}{8}, \frac{1}{16}$ … to infinite terms?

Note a = 1 and r = $\frac{1}{2}$; the sum of the series up to infinite terms = $\frac{1}{1-r} = \frac{1}{1-\frac{1}{2}} = \frac{1}{\frac{1}{2}} = 2$.

Practice Problems

1. What is the 9th term of the series -5, -2.5, 0, 2.5...?
   - A. -10
   - B. -17.5
   - C. 17.5
   - D. -20
   - E. 15

2. What is the 8th term of the series -2, 4, -8, 16...?
   - A. 256
   - B. -256
   - C. 512
   - D. -512
   - E. 1,024

3. How many terms are in the series 3, 10, 17, ..., 73?
   - A. 100
   - B. 90
   - C. 13
   - D. 12
   - E. 11

4. What is the sum of first 100 even numbers?
   - A. 10,000
   - B. 10,201
   - C. 9,999
   - D. 10,100
   - E. 10,002

5. What is the sum of the series 55, 60, 65, 70, ..., 235?
   - A. 5,220
   - B. 5,365
   - C. 5,300
   - D. 10,440
   - E. 10,730

6. Flowers in a pond double every day. On a Sunday, the pond is full. What day is pond one half full?
   - A. Friday
   - B. Thursday
   - C. Saturday
   - D. Wednesday
   - E. Monday

7. There are initially 243,000 Oxygen molecules in a chamber. Oxygen is depleted out of the chamber at a rate of $\frac{1}{3}$ the current amount per hour. How long does it take before 9,000 Oxygen molecules are in the chamber?
   - A. 1 hour
   - B. 2 hours
   - C. 3 hours
   - D. 4 hours
   - E. 10 hours

8. Find the sum of the infinite series $3, \frac{9}{5}, \frac{27}{25} ...,$
   - A. 7.5
   - B. 15
   - C. 3
   - D. 30
   - E. 75

9. How many terms are in the series if the initial term is 4, the common ratio is 3 and the sum of the terms is equal to 354,292?
   - A. 9
   - B. 13
   - C. 10
   - D. 11
   - E. 12

10. Abby deposits $1,200 in the bank. If the bank is paying 4% interest to be compounded annually, what is balance at the end of 4 years?
    - A. 1,350
    - B. 1,460
    - C. 1,392
    - D. 1,800
    - E. 1,404

11. What is the sum of the arithmetic series 4, Y, 16, Z, 28? There are 5 terms in the series. (Hint: the average = the middle term = 16).
    - A. Cannot be determined
    - B. 72
    - C. 80
    - D. 45
    - E. 105

12. What is the sum of the arithmetic series 2, 8, A, 20, B, C, D. There are 7 terms in the series. (Hint: the average = the middle term = 20).
    - A. Cannot be determined
    - B. 140
    - C. 98
    - D. 119
    - E. 120

13. What is the sum of 111, 211, 311, ..., 911?
    - A. 10,220
    - B. 9,198
    - C. 5,110
    - D. 4,599
    - E. Cannot be determined

Solutions to Practice Problems

1. ***Best answer is E.*** Note that it is an arithmetic series because -2×5+5=0-(-2.5)=2.5-0=2.5. The common difference, d = 2.5. The first term, a = -5. 9th term = $a + (n-1)d = -5 + (9-1) \times 2.5 = -5 + 20 = 15$.

2. ***Best answer is A.*** Note that the terms are changing sign alternatively. It is more likely that it is a geometric series. You can verify by noting that $\frac{4}{-2} = \frac{-8}{4} = \frac{16}{-8} = -2$. The first term is -2 and the common ratio is -2. 8th term = $ar^{n-1} = (-2)(-2)^{8-1} = 256$.

3. ***Best answer is E.*** The first term a = 3. The common difference, d = 10-3=17-10=7. The nth term = a + (n-1)d = 3+(n-1)7=73. Solving for n, n=11.

4. ***Best answer is D.*** First 100 even numbers are 2, 4, 6, ..., 198, 200. It is an arithmetic series with the first term = 2, the common difference = 2 and n = 100. Using the formula for sum of n terms = $n\frac{(\text{first term+last term})}{2} = 100\frac{(2+200)}{2} = 100 \times 101 = 10,100$.

5. ***Best answer is B.*** First we need to find out the number of terms, n. 55+(n-1)×5=235. Solving for n, n = 37. The sum of 37 terms = $n\frac{(\text{first term+last term})}{2} = 37\frac{(55+235)}{2} = 37 \times 145 = 5,365$.

6. ***Best answer is C.*** Day before Sunday, the pond should be half as full. Hence, on Saturday (day before Sunday) the pool is half full.

7. ***Best answer is C.*** Note that $9,000=243,000\times(\frac{1}{3})^n$; solving for n, n = 3.

8. ***Best answer is A.*** The common ratio, r = $\frac{(\frac{9}{5})}{3} = \frac{9}{5} \times \frac{1}{3} = \frac{3}{5}$; The sum of infinite terms = $\frac{\text{First term}}{1-r} = \frac{3}{1-\frac{3}{5}} = \frac{3}{(\frac{2}{5})} = 3 \times \frac{5}{2} = \frac{15}{2} = 7.5$.

9. ***Best answer is D.*** $354,292 = a\frac{r^n-1}{r-1}$; where a = 4 and r = 3; solving n, n = 11.

10. ***Best answer is E.*** Balance at the end of the 4 years = $1,200 \times (1 + 0.04)^4 = 1,200 \times 1.04^4 = 1,200 \times 1.1698 = \$1,404$.

11. ***Best answer is C.*** This is a trick question. You do not need to know Y and Z in the series. Because it is an arithmetic series, the middle number is the average. Hence, 16 is the average and there are 5 terms. The sum of the terms = 16×5=80.

12. ***Best answer is B.*** There are 7 terms and the average is 20. The sum of the terms = 7×20=140.

13. ***Best answer is D.*** Note that there are 9 terms in the series. The sum of 9 terms = $9\left(\frac{111+911}{2}\right) = 4,599$.

11. Permutations, Combinations and Probability

We will start this chapter with a simple example. Let us say, there are three islands A, B and C. A and B are connected by 3 bridges. B and C are connected by 4 bridges. See below:

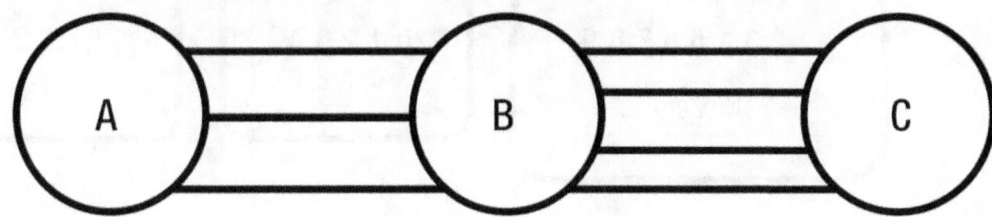

One would note that there are 3 ways to go from A to B and 4 ways to go from B to C. How many different ways are there to go from A to C? You can count these possibilities by hand and note that there are 12 different ways one can go from A to C.

Think of this as a "fill in the boxes" problem. There are two boxes. There are 3 ways to fill the first box and 4 ways to fill the second box. In total, there are 3×4=12 different ways to fill the boxes together.

This method can be applied to a variety of problems.

Example 1: If Sam, Josh, Jane, Mark are to take 4 seats at a dinner table, how many different ways can they be seated?

Think of this as a filling the boxes problems.

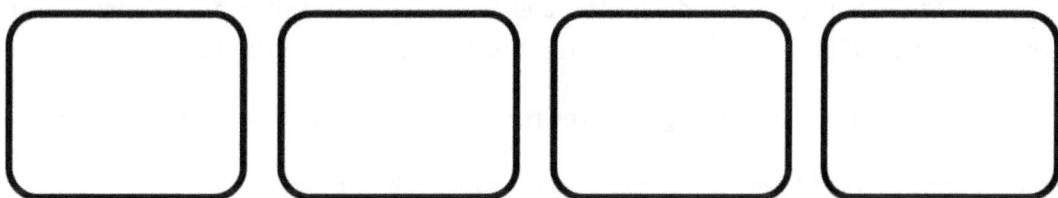

First box can be filled with any of the four choices. Once the first box is filled, there are only 3 choices. So, the second box can be filled with any of the three choices. Third box can be filled with 2 choices and last box can be filled with only one choice.

In total, there are 4×3×2×1 = 24 different ways. So, 4 guests can be seated in 24 different ways at the dinner table. 4×3×2×1 = 4! (it is also called 4 factorial).

Example 2: How many different "words" can be made from the letters L, G, A, U, S?

Note that there are 5 letters. Assuming repetitions are allowed (meaning same letter can be used more than once), the 5 boxes can be filled in 5×5×5×5×5 = 3,025 ways. So, one can make 3025 "words" with letters L, G, A, U, S.

If repetitions are not allowed, 5 boxes can be filled in 5×4×3×2×1 = 120 ways.

<u>Example 3:</u> How many 3 digit numbers can be made that end in 1, 4, or 7?

1, 2, 3, 4, 5, 6, 7, 8, 9 No Zero	0, 1, 2, 3, 4, 5, 6, 7, 8, 9	1, 4, 7

Think of this as a problem to fill 3 boxes. Note that there are 10 digits, 0, 1, 2,...9. The first box can be filled in 9 ways (not 10) because 0 cannot be the first digit. The second box has no restrictions and hence, it can be filled in 10 different ways. The third box can only be filled in three ways (1, 4 or 7). In total, there are 9×10×3 = 270 different three digit numbers that end in 1 or 4 or 7.

Permutations

Permutations involve arrangements. When talking about permutations, order is important. AB is not same as BA (because the order is important).

<u>Example 4:</u> There are 6 items being served at buffet. How many different ways can the items be arranged?

As discussed in the previous examples, the items can be arranged in 6×5×4×3×2×1 = 720 ways.

<u>Example 5:</u> Three couples Jack and Jill, Mark and Mary, and Beth and Bob are to be seated at a dinner table. How many different ways can they be seated if the couples are to sit together?

Note that the problem adds a <u>restriction</u>. The restriction is to keep the couples together.

Think of our problem as three big boxes that have two compartments each.

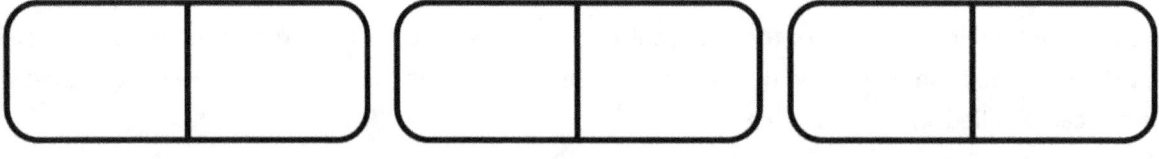

The three big boxes can be filled with couples in 3×2×1 = 6 different ways by placing J, M or B in the boxes. Once the big box is filled, note that the couples can be arranged in two ways 2×1 = 2 within the small compartments. Hence, there are 6×2×2×2 = 48 different ways to seat them.

<u>Example 6:</u> Three couples Jack and Jill, Mark and Mary, and Beth and Bob are to be seated at a dinner table. How many different ways can they be seated if Jack and Jill are to sit together?

Treat Jack and Jill as one unit. There are 5 entities to be seated at the table. They can be seated in 5! ways. Jack and Jill can be arranged in 2! ways. In total, there are 5!×2!=120×2=240 different ways to seat at a table so that Jack and Jill are seated together.

Example 7: Three couples Jack and Jill, Mark and Mary, and Beth and Bob are to be seated at a dinner table. How many different ways can they be seated so that men are seated together and women are seated together?

Now think that there are two big boxes and there are 3 smaller compartments within each box. The big boxes can be filled in two different ways with letters M (for men) and W (for women). Now, within each of the big boxes, there are 3×2×1=6 different ways to seat men and women.

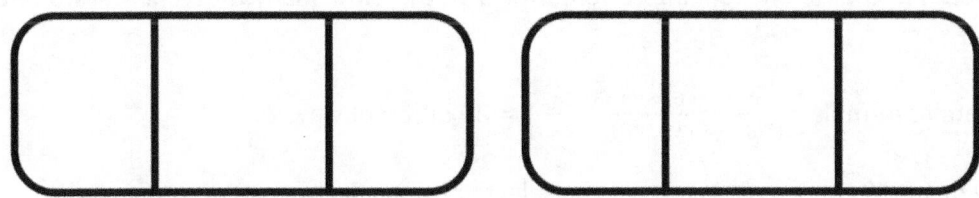

Hence, there are 2×6×6 = 72 different ways to seat both men and women together.

Example 8: Three couples Jack and Jill, Mark and Mary, and Beth and Bob are to be seated at a dinner table. How many different ways can they be seated if every man sits next to a woman (and vice versa)?

Number the boxes as 1, 2, 3, 4, 5, 6. Say, Boxes 1,3,5 can take men and boxes 2,4,6 can take women.

The boxes can be filled in (3×2×1)×(3×2×1) = 6×6=36 different ways.

However, we also have the option to fill boxes 1, 3, 5 with women and boxes 2,4,6 with men. This can also be accomplished in 36 ways. Hence, there are a total of 36+36 = 72 different ways to seat the guests.

General Permutations Formula

If there are "n" objects and if we have to arrange "r" (r<n) out of "n" objects in a sequence, we will have

$n.(n-1).(n-2)....r$ ways to arrange them. This can also be expressed as $\frac{n!}{(n-r)!}$.

Example 9: David has to pick 4 shoes from 10 shoes to place them in 4 racks. How many different ways David can pick the shoes?

Note that the first box can be filled in 10 different ways, second box in 9, third in 8 and fourth in 7.

Total number of ways to arrange the shoes = 10×9×8×7=5,040.

Combinations

Combination means choices or selections. <u>Order does not matter with combinations.</u> AB is same as BA. Assume there are three letters A, B and C and only two letters can be chosen.

We can select (AB/BA, AC/CA or BC/CB) 3 different ways.

General Combinations Formula

If there are "n" objects and if we have to choose "r" (r<n) out of "n" objects, we can choose "r" objects from "n" objects in $\frac{n!}{(n-r)!\, r!}$.

<u>Example 10</u>: If there are 6 different toppings offered for a pizza. How many different ways one can choose 3 toppings?

Using the combinations formula $= \frac{6!}{(3!3!)} = \frac{6\times5\times4\times3\times2\times1}{(1\times2\times3)(1\times2\times3)} = 20$ different ways.

<u>Example 11</u>: How many different ways 2 cards can be selected from a deck of cards?

Note that a single deck of cards has 52 cards. So, we are to select 2 cards from 52. Using the formula,

$$\frac{52!}{(50!2!)} = \frac{52\times51}{2\times1} = 1{,}326.$$

Probability

Probability deals with chance. Probability is a ratio. Probability is always between 0 and 1 (0 and 1 are included).

Probability $= \dfrac{\text{Number of favorable outcomes}}{\text{Number of total possible outcomes}}$

<u>Example 12</u>: What is the probability of heads on a single toss of a coin?

There are only two possibilities Heads or Tails. Favorable outcome is Heads (one outcome).

Hence, the probability of heads $= \frac{1}{2}$ = 0.5

<u>Example 13</u>: What is the probability of getting an odd number in a single roll of a die?

Note that a die has 6 faces (6 numbers 1 thru 6). 1, 3, 5 are odd numbers. The number of total outcomes is 6 and the number of favorable outcomes is 3.

Hence, the probability of getting an odd number $= \frac{3}{6} = \frac{1}{2}$.

<u>Example 14</u>: What is the probability of getting the same number when rolling a pair of dice?

Note that there are total 6×6=36 possibilities. The number of total possible outcomes = 36.

Favorable outcomes are (1,1); (2,2).. (6,6). There are 6 different favorable outcomes.

Hence, the probability = $\frac{6}{36} = \frac{1}{6}$.

Example 15: What is the probability to get an ace or a spade when a single card is drawn from a deck of cards?

There are 52 cards, 13 spades and 4 aces. However, one of the ace is a spade. Hence, there are 13+4-1=16 favorable outcomes.

The probability of getting a spade or an ace = $\frac{16}{52} = \frac{4}{13}$.

Example 16: There are 10 red fish, 11 green fish, 12 blue fish and 13 white fish in a pond. If one fish is taken from the pond at random, what is the probability of catching a green fish?

There are a total of 10+11+12+13=46 fish. There are 11 green fish.

The probability of catching a green fish = $\frac{11}{46}$.

Example 17: There are 10 red fish, 11 green fish, 12 blue fish and 13 white fish in a pond. If one fish is taken from the pond at random, what is the probability of not catching a red fish?

There are a total of 10+11+12+13=46 fish. There are 11+12+13=36 fish that are not red.

The probability of not catching a red fish = $\frac{36}{46} = \frac{18}{23}$.

Example 18: There are 10 red fish, 11 green fish, 12 blue fish and "X" white fish in a pond. If one fish is taken from the pond at random, what is the probability of catching a white fish?

There are a total of 10+11+12+X=33+X fish. There are "X" white fish.

The probability of catching a white fish = $\frac{X}{33+X}$.

Example 19: There are 10 red fish, 11 green fish, 12 blue fish and some white fish in a pond. If one fish is taken from the pond at random, the probability of getting a white fish is $\frac{1}{4}$. How many white fish are in the pond?

From the previous example, the probability of catching a white fish = $\frac{X}{33+X} = \frac{1}{4}$; solving for X, X = 11.

Example 20: A pond contains fish of two different colors. If one fish is taken at random without replacement, what is the maximum number (at most) of tries before one would find fish of matching color?

Because there are fish of two different colors, at the end of two tries one may not necessarily have a match. However, after the third try, the third one should match one of the first two because there are

80

only two possible colors. Hence, the answer is 3. In (at most) three tries, one must have fish of a matching color.

Practice Problems

1. A restaurant offers a pizza with 3 different crusts and 5 different toppings. How many different pizzas can be ordered with 1 crust and 2 toppings?

    A. 15   B. 30   C. 45   D. 60   E. 120

2. What is the probability of getting a Jack, Queen, King or an Ace upon a single draw from a deck of cards?

    A. $\frac{1}{2}$ B. 0   C. 1   D. $\frac{4}{13}$ E. $\frac{8}{13}$

3. How many different 3 digit numbers are divisible by 5?

    A. 90   B. 270   C. 999   D. 181   E. 180

4. What is the probability that a three digit number selected at random will be divisible by 5?

    A. $\frac{1}{2}$ B. $\frac{1}{5}$ C. $\frac{1}{10}$ D. $\frac{1}{4}$ E. $\frac{181}{900}$

5. Mary has 4 dresses, one of each color Green, Blue, Red and Purple. Each dress has a shirt and pant. How many different ways can she select a dress so that the shirt and pant are of the same color?

    A. 4   B. 8   C. 12   D. 16   E. 1

6. Mary has 4 dresses, one of each color Green, Blue, Red and Purple. Each dress has a shirt and pant. How many different ways can she select a dress so that the shirt and pant are of different color?

    A. 4   B. 8   C. 12   D. 16   E. 1

7. What is $\frac{(n-1)!}{n!}$?

    A. n   B. n-1   C. $\frac{1}{n}$ D. $\frac{1}{n-1}$ E. Cannot be determined

8. What is $\frac{n!}{(n-2)!}$?

    A. n   B. n-2   C. n-1   D. $n^2 - n$ E. Cannot be determined

9. How many different three digit numbers are possible whose tens digit is divisible by 2 and the units digit is 1 or 9?

    A. 180   B. 900   C. 450   D. 90   E. 45

10. A chess club has 10 players. How many different ways a team of 3 can be selected?

    A. 720   B. 24   C. 120   D. 240   E. 512

11. A safe has a 5 digit key. How many different ways a key can be made such that all digits are not the same and the key does not start with a zero or 9? (ex: not 11111, 22222, etc.)

    A. 79,992   B. 115   C. 24,192   D. 30,240   E. 9

12. A coffee shop offers 4 flavors of coffee. A customer can choose his/her coffee with whole milk or 2% milk or no milk. In addition, customer can choose from three different sizes tall, medium or small. How many different types of coffee can be ordered?

    A. 8   B. 12   C. 48   D. 24   E. 36

13. What is the probability that upon a roll of a pair of dice, the sum of two numbers is equal to 7?

    A. $\frac{7}{36}$ B. $\frac{5}{36}$ C. $\frac{1}{6}$ D. $\frac{1}{36}$ E. $\frac{1}{12}$

14. What is the probability that upon a roll of a pair of dice, the sum of two numbers is greater than 8?

    A. $\frac{7}{36}$ B. $\frac{5}{36}$ C. $\frac{5}{18}$ D. $\frac{1}{36}$ E. $\frac{1}{12}$

15. Jane and Jill went out for dinner with three other friends. What is the probability that Jane and Jill are seated together at a table?

    A. $\frac{2}{5}$ B. $\frac{5}{24}$ C. $\frac{1}{24}$ D. $\frac{3}{24}$ E. $\frac{3}{5}$

16. A pond contains fish of three different colors. If one fish is taken at random without replacement, what is the maximum number (at most) of tries before one would find fish of matching color?

    A. 2   B. 5   C. 3   D. 6   E. 4

17. How many diagonals are in a hexagon? (n=6)

    A. 6   B. 12   C. 15   D. 30   E. 9

18. How many diagonals are in a polygon with twelve sides? (n=12)

    A. 12   B. 132   C. 120   D. 54   E. 66

19. In a chess tournament, each player plays the other player twice. If there are 8 players in the tournament, how many total games are played?

    A. 56   B. 15   C. 28   D. 112   E. 120

20. In a round robin league, each team plays the other team once. Points are allocated as following: Win=4 points, Loss=0 points, Draw=1 point. If there are 8 teams in the tournament, what are the maximum points a team can score?

    A. 28   B. 32   C. 8   D. 7   E. 16

Solutions to Practice Problems

1. ***Best answer is B***. One crust can be selected from three choices in C_1^3 = 3 ways. Two toppings can be selected from 5 toppings in $C_2^5 = 10$ ways. In total, there are 3×10=30 different ways.

2. ***Best answer is D***. There are total of 52 cards in a single deck of cards. There are 4 Jacks, 4 Queens, 4 Kings and 4 Aces = a total of 16 favorable cards. The probability $= \frac{16}{52} = \frac{4}{13}$.

3. ***Best answer is E***. Think of this problem as filling three boxes. First box can be filled-in by any of the 9 digits 1 thru 9 (not a zero). Second box can be filled by 10 digits 0 thru 9. Third box can be filled by 0 or 5. Hence, there are a total of 9×10×2=180 different numbers.

4. ***Best answer is B***. There are 900 three digit numbers, 100 thru 999. As per problem 3, there are 180 three digit numbers that are divisible by 5. Hence, the probability is $\frac{180}{900} = \frac{1}{5}$. *SGK's Short Cut:* ha..ha..ha. One in every 5 numbers is divisible by 5; hence, the probability $= \frac{1}{5}$.

5. ***Best answer is A***. There are only 4 different ways. GG, BB, RR and PP.

6. ***Best answer is C***. Method 1: A shirt can be selected in 4 different ways and a pant in 4 different ways. Hence, when there are no restrictions, a dress can be selected in 4×4=16 ways. Out of those 16 dresses, 4 dresses (as per problem 5) match the color. Hence, 16-4=12 ways a dress can be selected without matching the color.

   SGK's Short Cut: A shirt can be selected in 4 different ways. Once a shirt is selected, there are only 3 pants that can be selected (because we cannot select the same color for the pant). Hence, there are 4×3=12 different ways of selecting a dress without matching the color.

7. ***Best answer is C***. $\frac{(n-1)!}{n!} = \frac{(n-1).(n-2).(n-3)...3.2.1}{n.(n-1).(n-2).(n-3)...3.2.1} = \frac{1}{n}$.

8. ***Best answer is D***. $\frac{n!}{(n-2)!} = \frac{n.(n-1).(n-2).(n-3)...3.2.1}{(n-2).(n-3)...3.2.1} = n(n-1) = n^2 - n$.

9. ***Best answer is D***. Think of it as filling three boxes. The first box, the hundreds digit can be filled by 1, 2, … thru 9, 9 different ways. The second box, tens digit can be filled by 0, 2, 4, 6 or 8, i.e., in 5 ways. Units digit is 1 or 9. Hence, there are 9×5×2=90 three digit numbers that can be formed to meet the criteria.

10. ***Best answer is C***. There are C_3^{10} ways of selecting the players. $C_3^{10} = \frac{10!}{(10-3)!3!} = \frac{10!}{7!3!} = \frac{10 \times 9 \times 8}{3 \times 2 \times 1} = 120$.

11. ***Best answer is A***. Number of 5 digit numbers that do not start with a zero or 9 = 8×10×10×10×10=80,000. Of these there are 8 different keys that are 11111, 22222,…,88888. Hence, possible combinations are 80,000-8=79,992.

12. ***Best answer is E***. There are 4×3×3=36 different ways.

13. ***Best answer is C***. Outcomes that add up to 7 are (1,6), (6,1), (2,5), (5,2), (4,3), (4,3) = 6 outcomes. Total possible outcomes are 6×6=36. The probability $= \frac{6}{36} = \frac{1}{6}$.

14. **Best answer is C**. Outcomes that add up to greater than 8 are (6,6), (6,5), (6,4), (6,3), (5,6), (5,5), (5,4), (4,6), (4,5), (3,6) = 10 outcomes. The probability $= \frac{10}{36} = \frac{5}{18}$.

15. **Best answer is A**. Treat Jane and Jill as one box with two compartments. Three friends and (Jane, and Jill) as one unit can be seated in 4! Ways. However, within the box, Jane and Jill can be seated in two different ways. Hence, there are 4!×2!=24×2=48 different ways they can be seated so that Jane and Jill are together. 5 guests (including Jane and Jill) can be seated in 5!=120 ways. Hence, the probability $= \frac{48}{120} = \frac{2}{5}$.

16. **Best answer is E**. After the first three tries, one might have fish of three different colors. However, on the fourth try one must have fish of matching color.

17. **Best answer is E**. n=6. One can draw a line between any two vertices. There are $C_2^6 = \frac{6!}{2!4!} = \frac{6 \times 5}{2 \times 1} = 15$ different lines that can be drawn between all 6 vertices. Out of these 15 lines, 6 lines represent the sides of the hexagon. Hence, the number of diagonals = 15-6=9. In general, a polygon with n sides will have "$C_2^n - n$" diagonals.

18. **Best answer is D**. n=12. One can draw a line between any two vertices. There are $C_2^{12} = \frac{12!}{2!10!} = \frac{12 \times 11}{2 \times 1} = 66$ different lines that can be drawn between all 12 vertices. Out of these 66 lines, 12 lines represent the sides of the polygon. Hence, the number of diagonals = 66-12=54. In general, a polygon with n sides will have "$C_2^n - n$" diagonals.

19. **Best answer is A**. n=8. By selecting two players from 8, one can form a game. There are $C_2^8 = \frac{8!}{2!6!} = \frac{8 \times 7}{2 \times 1} = 28$ games that can be played amongst 8 players. However, each player is playing the other twice. Hence, there are 28×2=56 games possible.

20. **Best answer is A**. Simple, each team plays 7 games accumulating at most 4×7=28 points.

12.   Set Theory

A set is defined as a collection of elements. For example, the set of natural numbers is {1, 2, 3, 4..}. A set of whole numbers is {0, 1, 2, 3,...}. A set can contain indefinite (infinite) elements or no elements at all. A set with no elements or zero elements is called a null set. Operations of a set are defined below.

Union of Two Sets

Union of two sets is also a set. Union contains all elements without repetition from each of the sets. Union is represented by symbol U.

Example1: Set A is {1, 3, 5, 7, 9} Set B is {3, 6, 9, 12}. What is A U B?

Then, A U B = {1, 3, 5, 6, 7, 9, 12}

Intersection of Two Sets

Intersection of two sets is also a set. Intersection contains elements that are common to the two sets. Intersection is represented by symbol ∩.

Example 2: Set A is {1, 3, 5, 7, 9} Set B is {3, 6, 9, 12}. What is A ∩ B?

Then, A ∩ B = {3, 9}.

Number of Elements in a Set

Example 3: Set A is {1, 3, 5, 7, 9} and Set B is {3, 6, 9, 12}. What are n(A) and n(B)?

Set A has 5 elements and Set B has 4 elements. We represent the number of elements for set A and B as follows: n(A) = 5 and n(B) = 4.

Relationship between Union and Intersection

For two sets A and B, n(AUB) = n(A) + n(B) − n(A∩B). Refer to the picture below. The picture is called a Venn Diagram.

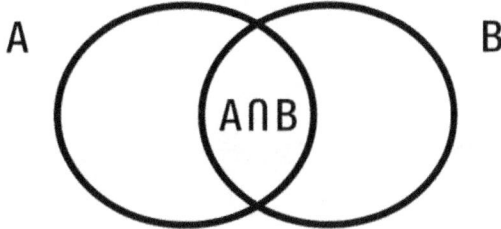

For three sets A, B and C, n(AUBUC) = n(A) + n(B) + n(C) + n(A∩B∩C) − n(A∩B) − n(B∩C) − n(A∩C).

Example 4:   There are 100 students in a class. All the students are required to take either Music or History or both. 45 students take Music and 65 students take History. How many students take both Music and History?

Say, A represents set of students who take Music and B represents set of students who take History.

$n(A) = 45$, $n(B)=65$, $n(A \cup B) = 100$.

The number of students who take both Music and History $= n(A \cap B) = n(A)+n(B)- n(A \cup B)=45+65-100 = 10$.

Example 5: There are 100 students in a class. 40 students take Music and 50 students take History. 10 students take both Music and History. How many students take neither Music nor History?

The number of students who take either Music or History $= n(A \cup B) = n(A)+n(B)- n(A \cap B)=40+50-10=80$.

Therefore, $100-80 = 20$ students take neither Music nor History.

Example 6: There are 500 students in a class. There are required to take one of the core classes English, Math or History. 200 students take Math, 200 students take English, 200 students take History. 50 students take both Math and English. 50 students take both English and History. 20 students take all three courses. How many students take Math and History?

Using the formula $n(A \cup B \cup C) = n(A) + n(B) + n(C) + n(A \cap B \cap C) - n(A \cap B) - n(B \cap C) - n(A \cap C)$,

The number of students who take either English or Math or History $= 500 = 200+200+200-50-50-n(A \cap C)+20$.

Solving for $n(A \cap C)$, $n(A \cap C)=20$.

The number of students taking Math and History $= 20$.

Practice Problems

1. A={1, 3, 5, 7} and B={2, 3, 4}. What is A∪B?
   A. {3}   B. {1, 2, 3, 4, 5, 7}   C. {1, 2, 3, 4, 5, 6, 7}   D. {1, 5, 7}
   E. {1, 2, 4, 5, 7}

2. A={1, 3, 5, 7} and B={2, 3, 4}. What is A∩B?
   A. {3}   B. {1, 2, 3, 4, 5, 7}   C. {1, 2, 3,4, 5, 6, 7}   D. {1, 5, 7}
   E. {1, 2, 4, 5, 7}

3. At a fast food restaurant, there are "X" number of customers in a day. 50 of them order both fries and soda. 20 of them order fries but not soda. 35 of them order soda but not fries. Assuming that there are no customers that did not order fries or soda, what is X?
   A. 200   B. 50   C. 55   D. 105   E. 5

4. From the picture, how many students did not participate in the survey?
   A. 200   B. 20   C. 70   D. 130   E. 75

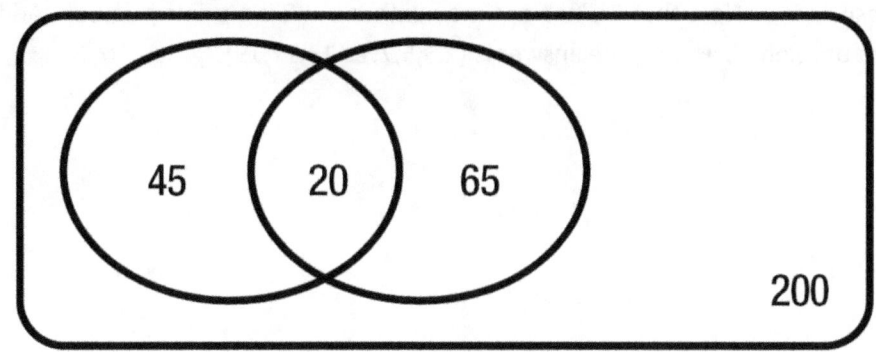

5. From the picture, how many students take at least two courses?
   A. 20   B. 5   C. 30   D. 25   E. 65

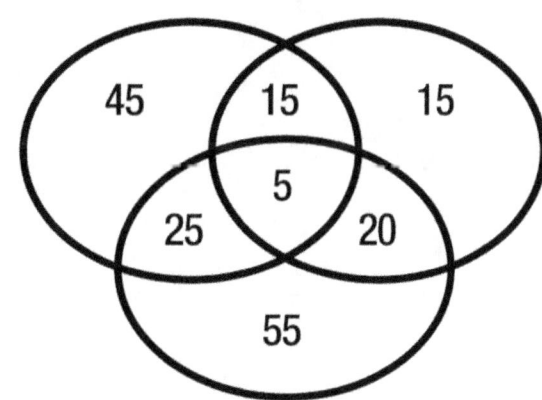

6. Which of the following represents a set of prime numbers less than 15?
   A. {1, 2, 3, 5, 7, 11, 13}   B. {2, 3, 5, 7, 11, 13} C. {3, 5, 7, 11, 13} D. {3, 5, 7, 9, 11, 13}
   E. {1, 2, 4, 5, 7}

7. Which of the following represents a set of prime numbers less than 20?
   A. {1,2,3,5,7,11,13,17,19} B. {2,3,5,7,11,13,17,19} C. {3,5,7,11,13,17,19} D. {3,5,7,9,11,13,17,19}
   E. {1,2,4,5,7,17,19}

Solutions to Practice Problems

1. **Best answer is B**. AUB contains all elements from both the sets. Hence, AUB = {1,2,3,4,5,7}.

2. **Best answer is A**. A∩B contains elements that are common to both the sets. A∩B={3}.

3. **Best answer is D**. The total number of customers = 20+50+35=105.

4. **Best answer is C**. From the picture, there are 200 students in the class. The number of students who participated in the survey = 45+20+65 =130. Therefore, 200-130=70 students did not participate in the survey.

5. **Best answer is E**. From the picture, the number of students taking at least two courses = 15+20+25+5=65.

6. **Best answer is B**. Note that 1 is not a prime number but 2 is a prime number. Hence, the answer is {2,3,5,7,11,13}.

7. **Best answer is B**. Note that 1 is not a prime number but 2 is a prime number. Add 17 and 19 to the list from problem 7. Hence, the answer is {2,3,5,7,11,13,17,19}.

13. Functions

A function is a mapping. It is a relationship between two variables. When we say y=f(x), we say, y is a function of x. In this case, x is called an independent variable and y is called a dependent variable. One can choose values of x at random or independently; whereas values of y cannot be chosen at random. Values of y are dictated by or determined by the relationship or the function f(x). x and y are also thought of as input and output respectively.

Domain and Range

Domain refers to values x can take. Range refers to values of y.

Example 1: y=5x+3. Note that x can take any values between negative infinity to positive infinity. So, can y. Hence, the domain and range equal to negative infinity to positive infinity.

Example 2: y=$\frac{1}{1-x}$. Note that when x=1, the denominator becomes 0. Division by zero is not defined. Hence, y is not defined when x=1. Hence, the domain is equal to negative infinity to 1 and 1 to positive infinity (but not including 1).

Example 3: $y = \frac{1}{x^2-4}$. Note that when x=2 or x=-2, the denominator becomes zero. Hence, the domain of this function does not include 2 or -2.

Definition of a Function

A function is a relationship where each input "x" has one and only one output "y".

In Example 1, where y = 5x + 3 is a function because each value of x yields only one value of y.

Example 4: $y^2 = x$ is not a function because each value of x yields two values of y. When x = 4, y= $\pm\sqrt{4}$=2 or -2.

Pictorial Representation of a Function

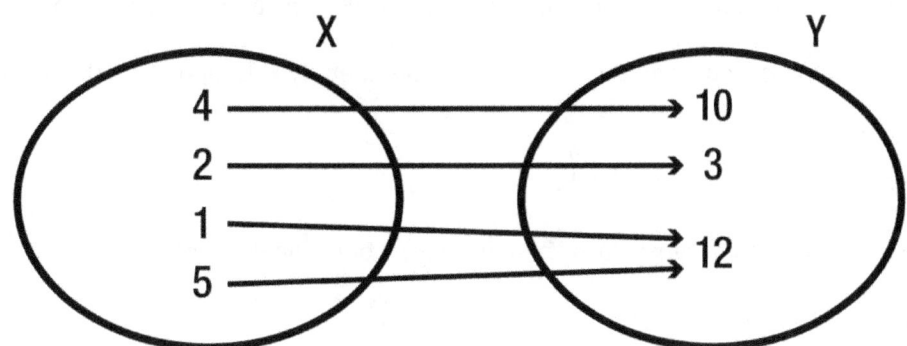

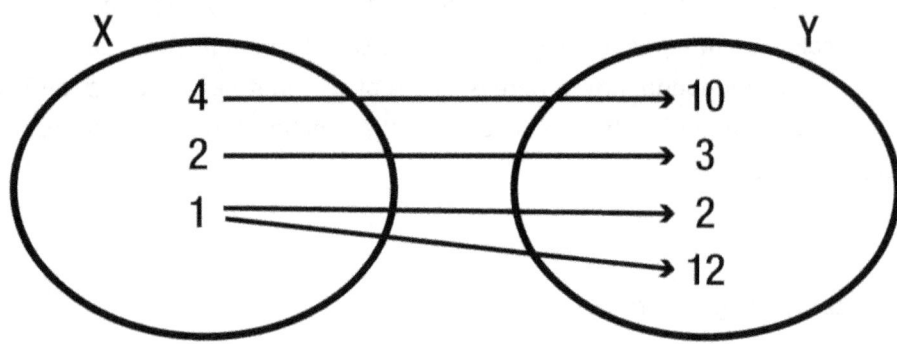

Inverse of a Function

Inverse of a function is represented as $f^{-1}(x)$. The inverse of a function is obtained by swapping the x and y variables. In example 1, swapping x and y variables, one would get x = 5y+3; solving for y, $y = \frac{x-3}{5}$. Note that in the original equation y = 5x+3, when x=1, y = 8. i.e., (1,8) satisfies the equation y = 5x+3. Note also that (8,1) satisfies the equation $y = \frac{x-3}{5}$. By definition, if (x,y) satisfies a function, then (y,x) should satisfy the inverse function. The inverse of a function is not necessarily a function.

Example 5: $y = x^2$. Each value of x yields only one value of y. Hence, f(x) is a function. Inverse is $y^2 = x$. As illustrated in the example 4, $y^2 = x$ is not a function.

Finding Inverse of a Function

Step 1: Test if given relationship is a function or not. i.e., each input yields one and only one output.

Step 2: Swap x and y variables.

Step 3: Solve for y.

Step 4: Test whether the inverse is a function or not. i.e., each input yields one and only one output.

Step 5: Verify your answer. If (x,y) is a point on the original function, then, (y,x) is a point on the inverse function.

Example 6: Find the inverse of function $f(x) = x^3 + 5$.

Step 1: $f(x) = y = x^3 + 5$ is a function because each value of x yields only one value of y.

Step 2: Interchange x and y. $x = y^3 + 5$.

Step 3: Solving for y, $y = \sqrt[3]{x - 5}$.

Step 4: Testing inverse function, each value of x yields only one value of y.

Hence, inverse function is $y = \sqrt[3]{x - 5}$.

Special Functions

The following types of problems stump best of the students. However, the problems are rather straightforward once you understand the concept.

<u>Example 7:</u> $\psi(X) = X^2 + 3X$; what is $\psi(5)$?

"ψ" is just a symbol representing a relationship. By comparison, note that X = 5 or you are asked to find out the value of the function when X = 5. Substituting X=5, $\psi(5) = 5^2 + 3 \times 5 = 25 + 15 = 40$.

<u>Example 8:</u> $X\psi Y = XY + X^2$; what is $2\psi 3$?

"ψ" is just symbol representing a relationship between X and Y. By comparison, note that X=2 and Y=3; hence, you just need to substitute 2 for X and 3 for Y. $2\psi 3 = 2 \times 3 + 2^2 = 6 + 4 = 10$.

Composite Function

Think of composite function as applying a function successively. For example f(g(x)) means taking output of g(x) and making it an input to the function f(x). This is best illustrated with examples.

<u>Example 9:</u> If $f(x) = x^2 + 1$ and $g(x) = 2x + 3$, what is $f\big(g(2)\big)$ and $g\big(f(2)\big)$?

First find out g(2) = 2×2+3 = 4+3=7. f(g(2)) = f(7) = $7^2 + 1 = 49 + 1 = 50$. f(2) = $2^2 + 1 = 4 + 1 = 5$. g(f(2)) = g(5) = 2×5+3 = 10+3 = 13. Note that f(g(x)) may not necessarily be equal to g(f(x)).

<u>Example 10:</u> $\psi(x)$ = 2x-3. What is $\psi(\psi(2))$?

Note that $\psi(2)$ = 2×2 -3 = 4-3=1. $\psi(\psi(2))$ = (1) = 2×1-3=2-3=-1

<u>Example 11:</u> $\psi(x) = ax^2 - 4x + 3$ and $\psi(3) = \psi(4)$, what is a?

$\psi(3) = \psi(4)$ means 9a-12+3=16a-16+3. Solving for a, 7a = 4 or a=$\frac{4}{7}$.

Practice Problems

1. If $f(x) = x^5 + 2$, what is inverse of $f(x)$?

   A. $\sqrt[3]{x+2}$ B. $\sqrt[3]{x-2}$ C. $\sqrt[5]{x-2}$ D. $\sqrt[5]{x+2}$ E. $x^5 - 2$

2. If $\psi(X) = \frac{X}{X+2}$, what is $\psi(5)$?

   A. $\frac{1}{5}$ B. $\frac{2}{7}$ C. $\frac{7}{2}$ D. $\frac{1}{3}$ E. $\frac{5}{7}$

3. If $\psi(X) = \frac{X}{X+2}$, for what value of X, $\psi(X) = X$?

   A. 1   B. 5   C. 2   D. -1   E. -2

4. If $\psi(X) = \frac{X}{X+2}$ and if $\psi(a) = 4$, what is a?

   A. $\frac{-3}{8}$ B. $\frac{3}{8}$ C. $\frac{-8}{3}$ D. $\frac{-2}{3}$ E. $\frac{8}{3}$

   For problems 5 thru 12: $f(x) = x^2 + 3$ and $g(x) = 2x + 3$.

5. What is $f\big(g(2)\big)$?

   A. 51   B. 0   C. 52   D. -51   E. -52

6. What is $g(f(2))$?

   A. 51   B. 0   C. 52   D. -51   E. 17

7. What is $f(g(x))$?

   A. $4x^2 + 12x + 9$ B. $4x^2 + 12x + 12$ C. $4x^2 + 12$ D. $4x^2 + 3$ E. $4x^2 + 12x$

8. What is $g(f(x))$?

   A. $4x^2 + 12x + 9$ B. $4x^2 + 12x + 12$ C. $4x^2 + 12$ D. $2x^2 + 9$ E. $4x^2 + 12x$

9. What is $f(f(x))$?

   A. $4x^2 + 12x + 9$ B. $4x^2 + 12x + 12$ C. $4x^2 + 12$ D. $4x^2 + 3$ E. $x^4 + 6x^2 + 12$

10. What is $g(g(x))$?

    A. $4x^2 + 12x + 9$ B. $4x^2 + 12x + 12$ C. $4x + 9$ D. $4x^2 + 3$ E. $4x^2 + 12x$

11. What is $g(g(1))$?

    A. 51   B. 13   C. 52   D. -51   E. 17

12. What is $f(f(2))$?

    A. 51   B. 13   C. 52   D. -51   E. 17

13. If $f(x) = 3x+4$, what is $f(f(3))$?

    A. 51   B. 13   C. 52   D. -51   E. 43

14. If $f(x) = 3x^3 + 5$, what is the inverse of $f(x)$?

    A. $y=x$ B. $y = \sqrt{\frac{x-5}{3}}$ C. $y = \sqrt[3]{\frac{x+5}{3}}$ D. $y = \sqrt[3]{\frac{x-5}{3}}$ E. Cannot be determined

Solutions to Practice Problems

1. **Best answer is C**. If $f(x) = x^5 + 2, y = x^5 + 2$, interchaning x and y, $x = y^5 + 2, y = \sqrt[5]{x - 2}$.

2. **Best answer is E**. Substituting X=5, $\psi(5) = \frac{5}{5+2} = \frac{5}{7}$.

3. **Best answer is D**. ψ (X)=X means, $\frac{X}{X+2} = X$; solving for X, X=-1.

4. **Best answer is C**. ψ (a)=4 means $\frac{a}{a+2} = 4$; solving for a, $a = \frac{-8}{3}$.

5. **Best answer is C**. g(2)=2×2+3=4+3=7. f(g(2))=f(7)=7×7+3=49+3=52.

6. **Best answer is E**. f(2)=2×2+3=4+3=7. g(f(2))=g(7)=2×7+3=14+3=17.

7. **Best answer is B**. $f(g(x))=f(2x+3)=(2x + 3)^2 + 3 = 4x^2 + 12x + 9 + 3 = 4x^2 + 12x + 12$.

8. **Best answer is D**. $g(f(x))=g(x^2 + 3) = 2(x^2 + 3) + 3 = 2x^2 + 9$.

9. **Best answer is E**. $f(f(x))=f(x^2 + 3) = (x^2 + 3)^2 + 3 = x^4 + 6x^2 + 9 + 3 = x^4 + 6x^2 + 12$.

10. **Best answer is C**. g(g(x))=g(2x+3)=2×(2x+3)+3=4x+6+3=4x+9.

11. **Best answer is B**. g(1)=2×1+3+2+3=5. g(g(1))=g(5)=2×5+3=10+3=13.

12. **Best answer is C**. f(2)=2×2+3+4+3=7. f(f(2))=f(7)=7×7+3=49+3=52.

13. **Best answer is E**. f(3)=3×3+4=9+4=13. f(f(3))=f(13)=3×13+4=39+4=43.

14. **Best answer is D**. $y = 3x^3 + 5$; interchaning x and y, $x = 3y^3 + 5; y = \sqrt[3]{\frac{x-5}{3}}$.

14. Matrices

A matrix is a lattice (rows and columns) of numbers. A matrix has row(s) and column(s).

$$A = \begin{bmatrix} 1 & 0 & 2 \\ 3 & -1 & 5 \end{bmatrix} \qquad B = \begin{bmatrix} 1 & 2 \\ 4 & 6 \\ 3 & 2 \end{bmatrix} \qquad C = \begin{bmatrix} 1 & 0 \\ 0 & 1 \end{bmatrix}$$

Matrix Size

Matrix size is indicated by r x c i.e., number of rows by number of columns. A 2×2 matrix has 2 rows and 2 columns. 1×3 matrix has 1 row and 3 columns. A 2×4 matrix has 2 rows and 4 columns. A n×n matrix has n rows and n columns.

Matrix Elements

Elements of a matrix are referenced by matrix with subscript row and column. For example, an element on the 2nd row and the 3rd column is referenced with index 2,3. In the example above, $A_{1,1} = 1$, $A_{1,2} = 0$, $A_{1,3} = 2$, $A_{2,1} = 3$, $A_{2,2} = -1$ and $A_{2,3} = 5$.

Matrix Addition and Subtraction

Matrix addition and subtraction are defined only for matrices of same size (i.e., matrices with same number of rows and columns). Addition and subtraction are conducted on the corresponding elements of each matrix. See Example below:

<u>Example 1:</u> If $A = \begin{bmatrix} 1 & 0 & 2 \\ 3 & -1 & 5 \end{bmatrix}$, $B = \begin{bmatrix} 1 & 2 \\ 4 & 6 \\ 3 & 2 \end{bmatrix}$ and $C = \begin{bmatrix} 2 & 1 & 2 \\ 7 & 4 & 0 \end{bmatrix}$, what is A+B and A+C?

Note that A+B is not defined because A and B are not of same size.

On the other hand, A+C= (Add corresponding Elements)= $\begin{bmatrix} 1+2 & 0+1 & 2+2 \\ 3+7 & -1+4 & 5+0 \end{bmatrix} = \begin{bmatrix} 3 & 1 & 4 \\ 10 & 3 & 5 \end{bmatrix}$.

Matrix Multiplication by a Constant Number

Multiplying a matrix by a scalar or a constant number is as simple as multiplying each element in the matrix by the constant.

<u>Example 2:</u> If $A = \begin{bmatrix} 1 & 0 & 2 \\ 3 & -1 & 5 \end{bmatrix}$, what is 3A?

Multiply each element by 3, $3A = \begin{bmatrix} 1 \times 3 & 0 \times 3 & 2 \times 3 \\ 3 \times 3 & -1 \times 3 & 5 \times 3 \end{bmatrix} = \begin{bmatrix} 3 & 0 & 6 \\ 9 & -3 & 15 \end{bmatrix}$.

Matrix Multiplication

Matrix multiplication is defined only for certain matrices. Matrix multiplication of matrix A of size m×n is defined with matrix B of size n×k resulting in a new matrix of size m×k. That means, AB is defined if and only if number of columns in matrix A = number of rows in matrix B.

Matrix multiplication of two matrices A and B is performed as follows:

- Step 1: Take the first row of A and the first column of B.

  - In the example 3 below, the first row of A is $[1 \quad 0 \quad 2]$ and the first column of B is $\begin{bmatrix} 1 \\ 4 \\ 3 \end{bmatrix}$

- Step 2: Multiply the corresponding elements from the first row of A and the elements from the first column of B. Add the results from the multiplication to obtain the first element of new matrix AB.

  - To make it easy, convert the first row of A into a column. i.e., the first row of A is $\begin{bmatrix} 1 \\ 0 \\ 2 \end{bmatrix}$

  - Multiplying corresponding elements means $\begin{bmatrix} 1 \\ 0 \\ 2 \end{bmatrix} \longrightarrow \begin{bmatrix} 1 \\ 4 \\ 3 \end{bmatrix}$; 1×1, 0×4 and 2×3 yields 1, 0 and 6.
  Add them to obtain 7.

- Step 3: Take the first row of A and the second column of B.

  - In the example 3 below, the first row of A is $[1 \quad 0 \quad 2]$ and the first column of B is $\begin{bmatrix} 2 \\ 6 \\ 2 \end{bmatrix}$

- Step 4: Multiply the corresponding elements from the first row of A and the elements from the second column of B. Add the results from the multiplication to obtain the second element of new matrix AB.

  - To make it easy, convert the first row of A into a column. i.e., the first row of A is $\begin{bmatrix} 1 \\ 0 \\ 2 \end{bmatrix}$

  - Multiplying corresponding elements means $\begin{bmatrix} 1 \\ 0 \\ 2 \end{bmatrix} \longrightarrow \begin{bmatrix} 2 \\ 6 \\ 2 \end{bmatrix}$; 1×2, 0×6 and 2×2 yields 2, 0 and 4.
  Add them to obtain 6.

- Repeat the process by taking each row from matrix A and each column from matrix B.
  - To make it easy, convert the second row of A into a column. i.e., the second row of A is $\begin{bmatrix} 3 \\ -1 \\ 5 \end{bmatrix}$

  - Multiplying corresponding elements means $\begin{bmatrix} 3 \\ -1 \\ 5 \end{bmatrix} \longrightarrow \begin{bmatrix} 1 \\ 4 \\ 3 \end{bmatrix}$; 3×1, -1×4 and 5×3 yields 3, -4 and 15. Add them to obtain 14.

o Multiplying corresponding elements means $\begin{bmatrix} 3 \\ -1 \\ 5 \end{bmatrix} \rightarrow \begin{bmatrix} 2 \\ 6 \\ 2 \end{bmatrix}$; 3×2, -1×6 and 5×2 yields 6, -6 and 10. Add them to obtain 10.

<u>Example 3:</u> If A = $\begin{bmatrix} 1 & 0 & 2 \\ 3 & -1 & 5 \end{bmatrix}$, B = $\begin{bmatrix} 1 & 2 \\ 4 & 6 \\ 3 & 2 \end{bmatrix}$, what is AB and BA?

Matrix A has 3 columns and Matrix B has 3 rows. Hence, multiplication AB is defined.

AB = $\begin{bmatrix} 1 \times 1 + 0 \times 4 + 2 \times 3 & 1 \times 2 + 0 \times 6 + 2 \times 2 \\ 3 \times 1 + (-1 \times 4) + 5 \times 3 & 3 \times 2 + (-1 \times 6) + 5 \times 2 \end{bmatrix} = \begin{bmatrix} 1+0+6 & 2+0+4 \\ 3-4+15 & 6-6+10 \end{bmatrix} = \begin{bmatrix} 7 & 6 \\ 14 & 10 \end{bmatrix}$.

BA = $\begin{bmatrix} 1 \times 1 + 2 \times 3 & 1 \times 0 + 2 \times -1 & 1 \times 2 + 2 \times 5 \\ 4 \times 1 + 6 \times 3 & 4 \times 0 + 6 \times -1 & 4 \times 2 + 6 \times 5 \\ 3 \times 1 + 2 \times 3 & 3 \times 0 + 2 \times -1 & 3 \times 2 + 2 \times 5 \end{bmatrix} = \begin{bmatrix} 1+6 & 0-2 & 2+10 \\ 4+18 & 0-6 & 8+30 \\ 3+6 & 0-2 & 6+10 \end{bmatrix} = \begin{bmatrix} 7 & -2 & 12 \\ 22 & -6 & 38 \\ 9 & -2 & 16 \end{bmatrix}$

AB is a 2x2 matrix and BA is a 3x3 matrix. Note that AB is not necessarily the same as BA. In addition, if AB is defined that does not mean BA is defined. AB and BA are both defined if and only if matrix A is of the size "m by n" and matrix B is of the size "n by m". Then, AB will be of the size "m by m" and BA will be of the size "n by n".

Matrix Determinant

Determinant of a 2×2 matrix $\begin{vmatrix} a & b \\ c & d \end{vmatrix}$ is defined as ad - bc.

<u>Example 4:</u> What is the determinant of the matrix $\begin{bmatrix} 7 & 6 \\ 14 & 10 \end{bmatrix}$?

By definition, the determinant=7×10-6×14=70-84=-14.

Matrices – Applications to Solving Simultaneous Equations

Let us consider simultaneous equations: 2x+3y=8 and 3x+4y=11.

The above two equations can be expressed in the matrix form as indicated below.

$$\begin{bmatrix} 2 & 3 \\ 3 & 4 \end{bmatrix} \begin{bmatrix} x \\ y \end{bmatrix} = \begin{bmatrix} 8 \\ 11 \end{bmatrix}$$

Two simultaneous equations in two variables have a unique solution if and only if determinant of the coefficient matrix is non-zero. The determinant of matrix $\begin{bmatrix} 2 & 3 \\ 3 & 4 \end{bmatrix} = 2 \times 4 - 3 \times 3 = 8 - 9 = -1$. Hence, the set of equations has a unique solution.

<u>Example 5:</u> If $\begin{bmatrix} 2 & 0 \\ 0 & 4 \end{bmatrix} \begin{bmatrix} x \\ y \end{bmatrix} = \begin{bmatrix} 8 \\ 12 \end{bmatrix}$, find x and y.

Using matrix multiplication, 2×x+0×y=8 and 0×x+4×y=12; 2x=8 and 4y=12; x=4 and y=3.

Example 6: If 3 pencils and 4 books cost $2 and 5 pencils and 1 book costs $2.5, represent it in a matrix form.

$$\begin{bmatrix} 3 & 4 \\ 5 & 1 \end{bmatrix} \begin{bmatrix} x \\ y \end{bmatrix} = \begin{bmatrix} 2 \\ 2.5 \end{bmatrix}$$

Practice Problems

1. If $A = \begin{bmatrix} 1 & 2 & 2 \\ 3 & 0 & 5 \end{bmatrix}$, $B = \begin{bmatrix} 1 & 2 \\ -1 & 3 \\ 2 & 5 \end{bmatrix}$, what is A+B?

    A. $\begin{bmatrix} 2 & 4 & 2 \\ 3 & 2 & 10 \end{bmatrix}$ B. $\begin{bmatrix} 2 & 4 \\ 2 & 3 \\ 2 & 10 \end{bmatrix}$ C. $\begin{bmatrix} 1 & 2 \\ -1 & 3 \\ 2 & 5 \end{bmatrix}$ D. $\begin{bmatrix} 1 & 2 & 2 \\ 3 & 0 & 5 \end{bmatrix}$ E. Not Defined

2. What is AB?

    A. $\begin{bmatrix} 2 & 4 & 2 \\ 3 & 2 & 10 \end{bmatrix}$ B. $\begin{bmatrix} 2 & 4 \\ 2 & 3 \\ 2 & 10 \end{bmatrix}$ C. $\begin{bmatrix} 3 & 18 \\ 13 & 31 \end{bmatrix}$ D. $\begin{bmatrix} 1 & 2 & 2 \\ 3 & 0 & 5 \end{bmatrix}$ E. Not Defined

3. What is 5A?

    A. $\begin{bmatrix} 5 & 10 & 10 \\ 15 & 0 & 25 \end{bmatrix}$ B. $\begin{bmatrix} 2 & 4 \\ 2 & 3 \\ 2 & 10 \end{bmatrix}$ C. $\begin{bmatrix} 3 & 18 \\ 13 & 31 \end{bmatrix}$ D. $\begin{bmatrix} 1 & 2 & 2 \\ 3 & 0 & 5 \end{bmatrix}$ E. Not Defined

4. What is 4A+3B?

    A. $\begin{bmatrix} 2 & 4 & 2 \\ 3 & 2 & 10 \end{bmatrix}$ B. $\begin{bmatrix} 2 & 4 \\ 2 & 3 \\ 2 & 10 \end{bmatrix}$ C. $\begin{bmatrix} 3 & 18 \\ 13 & 31 \end{bmatrix}$ D. $\begin{bmatrix} 1 & 2 & 2 \\ 3 & 0 & 5 \end{bmatrix}$ E. Not Defined

5. Given $\begin{bmatrix} 1 & 0 \\ 0 & 5 \end{bmatrix} \begin{bmatrix} x \\ y \end{bmatrix} = \begin{bmatrix} 4 \\ 10 \end{bmatrix}$, what is x?

    A. 3   B. 2   C. 4   D. 1   E. 0

6. Given $\begin{bmatrix} 1 & 0 \\ 0 & 5 \end{bmatrix} \begin{bmatrix} x \\ y \end{bmatrix} = \begin{bmatrix} 4 \\ 10 \end{bmatrix}$, what is y-x?

    A. 3   B. 2   C. 4   D. 1   E. -2

7. $\begin{bmatrix} 2 & 3 \\ 1 & 5 \end{bmatrix} \begin{bmatrix} x \\ y \end{bmatrix} = \begin{bmatrix} 4 \\ 7 \end{bmatrix}$ and if x represents number of men and y represents number of women in a certain class, which of the following represents a relationship between the number of men and women?

    A. 2x+3y=7   B. 2x+3y=4   C. x+5y=4   D. x+y=4   E. x=4

8. The determinant of a 2x2 matrix $\begin{vmatrix} a & b \\ c & d \end{vmatrix}$ is defined as ad-bc. What is the determinant of $\begin{bmatrix} 2 & 3 \\ 1 & 5 \end{bmatrix}$?

    A. 3   B. 2   C. 4   D. 1   E. 7

9. If the determinant of $\begin{bmatrix} x & 3 \\ 1 & 5 \end{bmatrix}$ is equal to 2, what is x?

    A. 3   B. 2   C. 4   D. 1   E. -2

10. If the determinant of $\begin{bmatrix} x & 8 \\ 2 & x \end{bmatrix}$ is equal to 0, what is x?

    A. 3   B. 2   C. 4   D. -2   E. 8

Solutions to Practice Problems

1. **Best answer is E**. Because A and B are not of same size, A+B or A-B are not defined.

2. **Best answer is C**. AB is defined. $AB = \begin{bmatrix} 1 \times 1 + (2 \times -1) + 2 \times 2 & 1 \times 2 + 2 \times 3 + 2 \times 5 \\ 3 \times 1 + (0 \times -1) + 5 \times 2 & 3 \times 2 + 0 \times 3 + 5 \times 5 \end{bmatrix} = \begin{bmatrix} 1 - 2 + 4 & 2 + 6 + 10 \\ 3 + 0 + 10 & 6 + 0 + 25 \end{bmatrix} = \begin{bmatrix} 3 & 18 \\ 13 & 31 \end{bmatrix}$.

3. **Best answer is A**. Multiply each element of A with 5 to get $5A = \begin{bmatrix} 5 \times 1 & 5 \times 2 & 5 \times 2 \\ 5 \times 3 & 5 \times 0 & 5 \times 5 \end{bmatrix} = \begin{bmatrix} 5 & 10 & 10 \\ 15 & 0 & 25 \end{bmatrix}$.

4. **Best answer is E**. Because A+B is not defined, 4A+3B is not defined either.

5. **Best answer is C**. Matrix multiplication gives 1×x+0×y=4; which means x=4.

6. **Best answer is E**. 1×x+0×y=4 and 0×x+5×y=10; which means x=4 and $y=\frac{10}{5}=2$. y-x=2-4=-2.

7. **Best answer is B**. Matrix multiplication gives 2x+3y=4 and x+5y=7.

8. **Best answer is E**. By definition, determinant=2×5-1×3=10-3=7.

9. **Best answer is D**. By definition, determinant = x×5-1×3=2 or 5x-3=2; solving for x, x=1.

10. **Best answer is C**. By definition, determinant x.x-2×8=0 or x^2=16 or x=4 or -4.

15.   Complex Numbers

Let us start with an example. How do you solve the equation $x^2 + 25 = 0$. $x^2 = -25$ and $x = \sqrt{-25}$. There is no real number to represent $\sqrt{-25}$. To simplify, we write $\sqrt{-1} = i$. Then, we can write $\sqrt{-25} = \sqrt{25} \times \sqrt{-1} = \pm 5i$. 5i is an imaginary number or a complex number.

General Form of a Complex Number

In general, a complex number can be represented as "a+bi", where a and b are real numbers (b≠0). "a" is called the real part and "b" is called the imaginary part.

Adding and Subtracting Complex Numbers

Complex numbers are added and subtracted by operating on the "real" and "imaginary" parts individually. For example, to add (3+4i) and (2+5i) = (add the real parts) + (add the imaginary parts)

(3+4i)+(2+5i)=(3+2) + (4+5) i = 5+9i.

(2+i)-(4-3i) = (2-4)+(1-(-3))i = -2+4i.

Complex Conjugate

A complex conjugate of a complex number is obtained by flipping the sign of the imaginary part. For example, complex conjugate of 3+4i is 3-4i. Complex conjugate of 2+5i is 2-5i. Complex conjugate of 9i is -9i.

Multiplication of Complex Numbers

Complex numbers are multiplied using foiling (or associative law).

2×3i = 6i.

-3×4i=-12i.

(2+3i)×(3+4i) = 2(3+4i)+3i(3+4i)=(2×3)+(2×4i) (3i×3)+(3i×4i)=6+8i+9i+12i^2 =6+17i-12 =-6+17i.

Note that i^2=-1 because $\sqrt{-1} \times \sqrt{-1} = -1$.

Properties of a Complex Conjugate

- When you add a complex number and its complex conjugate, you get a real number.
  For example, (3+4i)+(3-4i)=(3+3)+(4i-4i)=6 = 2×3=2 (real part).
- When you multiply a complex number and its complex conjugate, you get a real number.
  For example, (3+4i)×(3-4i) = (3×3) + (3×4i)-(3×4i)-(4i×4i)=9+12i-12i-16i^2=9-16×(-1) = 9+16=25.
  Note also that $3^2 + 4^2$ = 9+16 = 25. This is called the square of the magnitude.
  (2+5i)(2-5i) can be written as $2^2 + 5^2$ = 2×2+5×5 = 4+25=29 without doing the multiplication.
  (1+7i)(1-7i)= $1^2 + 7^2$=1+49=50.

101

(6+2i) (6-2i)= $6^2 + 2^2$ =36+4=40.

- Complex conjugate is useful when performing division with a complex number.

Division with a Complex Number

How do we calculate $\frac{1+2i}{3+4i}$?

Although it seems difficult to perform a division with "3+4i", it is made possible by complex conjugate.

$$\frac{1+2i}{3+4i} = \frac{1+2i}{3+4i} \times \frac{3-4i}{3-4i} = \frac{(1+2i)(3-4i)}{3\times3+4\times4} = \frac{3+6i-4i-8i^2}{25} = \frac{3+2i+8}{25} = \frac{11+2i}{25} = \frac{11}{25} + \frac{2}{25}i.$$

We multiplied both the numerator and the denominator by 3-4i, which is the complex conjugate of 3+4i. Note that a complex number multiplied by its complex conjugate results in a real number.

We also note that (3+4i)×(3-4i) = 3×3+4×4 = 9+16=25.

Calculating higher powers of i

Note the following:

$i^2 = i \times i = \sqrt{-1} \times \sqrt{-1} = -1.$

$i^3 = i^2 \times i = -1 \times i = -i.$

$i^4 = i^2 \times i^2 = -1 \times -1 = 1.$ This is a very important finding.

$i^5 = i^4 \times i = 1 \times i = i.$

$i^6 = i^4 \times i^2 = 1 \times i^2 = -1.$

$i^7 = i^4 \times i^3 = 1 \times -i = -i.$

$i^8 = i^4 \times i^4 = 1 \times 1 = 1.$

Note that the answer repeats after every 4th time. What that means is, you only need to know the first four. For example, to find i^{101}, divide 101 by 4. Note that 1 is the remainder. Hence, $i^{101} = i^1=i.$

To find i^{202}, divide 202 by 4. Note that 2 is the remainder. Hence, $i^{202}=i^2=-1.$

To find, i^{787}, divide 787 by 4. Note that 3 is the remainder. Hence, $i^{787}=i^3=-i.$

To find, i^{800}, divide 800 by 4. Note that 0 is the remainder. Hence, $i^{800} = i^4=1.$

Practice Problems

1. What is complex conjugate of 2+9i
   - A. 2
   - B. 9
   - C. 2+9i
   - D. 2-9i
   - E. -9i

2. What is the sum of 2+9i and its complex conjugate?
   - A. 4+18i
   - B. 4-18i
   - C. 4
   - D. 18i
   - E. -18i

3. What is the product of 2+9i and its complex conjugate?
   - A. 4+18i
   - B. 4-18i
   - C. 17+36i
   - D. 85
   - E. 0

4. Multiply (1+2i) and (3+4i).
   - A. -5+10i
   - B. 3+6i
   - C. 11+10i
   - D. 4+6i
   - E. 11

5. Add (1+2i) and (3+4i).
   - A. -5+10i
   - B. 3+6i
   - C. 11+10i
   - D. 4-6i
   - E. 4+6i

6. What is $\frac{1+2i}{3-4i}$ equal to?
   - A. -5+10i
   - B. 3+6i
   - C. 11+10i
   - D. 4+6i
   - E. $\frac{-5}{25} + \frac{10}{25}i$

7. What is $i^3 \times i^7$ equal to?
   - A. i
   - B. -1
   - C. 1
   - D. −i
   - E. 0

8. What is i^{203} equal to ?
   - A. i
   - B. -1
   - C. 1
   - D. −i
   - E. 0

9. What is i^{991} equal to?
   - A. i
   - B. -1
   - C. 1
   - D. −i
   - E. 0

10. What is $\frac{1}{2+3i}$ equal to?
    - A. -5+10i
    - B. 3+6i
    - C. 11+10i
    - D. 4+6i
    - E. $\frac{2}{13} - \frac{3}{13}i$

Solutions to Practice Problems

1. **Best answer is D.** Complex conjugate is obtained by flipping the sign of the complex part. So, it is 2-9i.

2. **Best answer is C.** Note that the sum of a complex number and its conjugate is always a real number. Choice C is the only real number. 2+9i+(2-9i)=4.

3. **Best answer is D.** Product of a complex number with its conjugate =square of the amplitude=2×2+9×9=4+81=85. _SGK's Short Cut:_ The answer should be a real number. Choice E is obviously wrong. Hence, you can pick Choice D without really solving the problem.

4. **Best answer is A.** $(1 + 2i)(3 + 4i) = 1 \times 3 + 1 \times 4i + 2i \times 3 + 2i \times 4i = 3 + 10i + 8i^2 = 3 - 8 + 10i = -5 + 10i$.

5. **Best answer is E.** $(1 + 2i) + (3 + 4i) = 1 + 3 + (2 + 4)i = 4 + 6i$. Add the real and imaginary parts together.

6. **Best answer is E.** Choice E is the only choice with a division by 25 (3×3+4×4=25). You can pick this choice without doing all the math. $\frac{1+2i}{3-4i} = \frac{1+2i}{3+4i} \times \frac{3+4i}{3+4i} = \frac{(1+2i)(3+4i)}{3\times3+4\times4} = \frac{3+6i+4i+8i^2}{9+16} = \frac{3-8+10i}{25} = \frac{-5+10=2i}{25} = \frac{-5}{25} + \frac{10}{25}i$.

7. **Best answer is B.** $i^3 \times i^7 = i^{10} = i^4 \times i^4 \times i^2 = 1 \times 1 \times (-1) = -1$.

8. **Best answer is D.** $i^{203} = i^{200} \times i^3 = 1 \times i^3 = -i$.

9. **Best answer is A.** $i^{991} = i^{900} \times i^1 = 1 \times i = i$.

10. **Best answer is E.** $\frac{1}{2+3i} = \frac{1}{2+3i} \cdot \frac{2-3i}{2-3i} = \frac{2-3i}{2\times2+3\times3} = \frac{2-3i}{4+9} = \frac{2-3i}{13} = \frac{2}{13} - \frac{3}{13}i$.

16.   Logarithms

During the time when there were no calculators, logarithmic tables were used instead. When we look at b^x, it is important to note that "b" is called the base and "x" is called the exponent. Logarithms are like inverse operations for exponents.

For example, if $y = b^x$, by definition, $\log_b y = x$; this definition forms the basis for the rest of the chapter. Because the base can be any number, in our example, "b" can be 2, 5, 10, or any other number. When the base is 10, it is omitted. For example log x is same as log x to the base 10. log 5 = $\log_{10} 5$.

In logarithms, there is a special base, indicated by "e". "e" is a number as defined by $e = 1 + 1 + \frac{1}{2!} + \frac{1}{3!} + \frac{1}{4!} + \cdots$ (up to infinite numbers) ≈ 2.7183. When base is e, the logarithm is written as ln. ln means to the base "e". ln 2 = ln 2 to the base e. It is also called the natural logarithm.

Rules of Addition and Subtraction

Remember from exponents: when the base is same, add (or subtract) exponents.

When exponents are the same, multiply the bases.

We use these principles to define the rules for addition and subtraction of logarithms.

$\log a + \log b = \log(ab)$.

$\log a - \log b = \log(\frac{a}{b})$.

$\log a^y = y \log a$.

$\log_b a = \frac{\log a}{\log b}$.

Example 1: $\log_{10} 100 = 2$; $\log_{10} 1000 = 3$ because $10^2 = 100$ and $10^3 = 1000$.

Example 2: $\log 3 + \log 5 = \log(3 \times 5) = \log(15)$.

Example 3: $\log 10 - \log 2 = \log\left(\frac{10}{2}\right) = \log(5)$.

Example 4: $\log 10 = 1$ because $10^1 = 10$. Using the formula $\log_b a = \frac{\log a}{\log b}$, it can be derived that $\log_a a = \frac{\log a}{\log a} = 1$.

Example 5: $\log 1 = 0$. Using the formula $\log a - \log b = \log(\frac{a}{b})$, it can be derived that $\log a - \log a = \log\left(\frac{a}{a}\right)$; $\log 1 = 0$.

Example 6: $\frac{\log 20}{\log 5} = \log_5 20$.

Example 7: $\log_2 3 \times \log_3 4 = \log_2 4$.

Using the formula $\log_b a = \frac{\log a}{\log b}$, $\log_2 3 = \frac{\log 3}{\log 2}$ and $\log_3 4 = \frac{\log 4}{\log 3}$.

Hence, $\log_2 3 \times \log_3 4 = \frac{\log 3}{\log 2} \times \frac{\log 4}{\log 3} = \frac{\log 4}{\log 2} = \log_2 4 = \log_2 2^2 = 2\log_2 2 = 2 \times 1 = 2$.

Practice Problems

1. What is $\log 7 + \log 5$?

    A. $\log 12$ B. $\log 35$ C. $\log(\frac{7}{5})$ D. 35   E. $\log 2$

2. What is $\log 7 - \log 5$?

    A. $\log 12$ B. $\log 35$ C. $\log\left(\frac{7}{5}\right)$ D. 35   E. $\log 2$

3. What is $\log 12 + \log 6 - \log 3$?

    A. $\log 15$ B. $\log 35$ C. $\log 24$ D. 24   E. $\log 2$

4. What is $\log 20 + \log 5$?

    A. $\log 25$ B. $\log 35$ C. $\log 24$ D. 2   E. $\log 2$

5. What is $\log_3 45 - \log_3 5$?

    A. $\log 25$ B. $\log 35$ C. $\log 24$ D. 3   E. 2

6. If $4^{2x+8} = 8^{5x}$, what is x?

    A. $\frac{8}{3}$ B. 1   C. 3   D. $\frac{16}{11}$ E. $\frac{11}{16}$

7. If $9^{2x-8} = 243^x$, what is x?

    A. $\frac{8}{3}$ B. -16   C. -8   D. $\frac{16}{11}$ E. $\frac{11}{16}$

8. Simplify $\ln e^5$

    A. 5   B. 2.88   C. 3   D. 10   E. $\ln 5$

9. Simplify $\ln e^5 - \ln e$

    A. 5   B. 2.88   C. 4   D. 10   E. $\ln 5$

10. What is $\frac{\log 15}{\log 5}$?

    A. $\log 3$ B. $\log 10$ C. $\log_5 15$ D. 3   E. 2

11. What is $\log_4 25 \times \log_3 4$?

    A. $\log 25$ B. $\log 100$ C. $2\log_3 5$ D. 3   E. Cannot be determined

Solutions to Practice Problems

1. **Best answer is B**. Using the addition rule, $\log 7 + \log 5 = \log(7 \times 5) = \log 35$.

2. **Best answer is C**. Using the subtraction rule, $\log 7 - \log 5 = \log\left(\frac{7}{5}\right)$.

3. **Best answer is C**. Using the addition and subtraction rule, $\log 12 + \log 6 - \log 3 =$
    $\log\left(\frac{12 \times 6}{3}\right) = \log 24$.

4. **Best answer is D**. Using the addition rule, $\log 20 + \log 5 = \log(20 \times 5) = \log 100 = \log_{10} 10^2$
    $= 2\log_{10} 10 = 2$.

    Remember $\log$ means to the base 10 by default and $\log_{10} 10 = 1$.

5. **Best answer is E**.

    Using the subtraction rule, $\log_3 45 - \log_3 5 = \log_3\left(\frac{45}{5}\right) = \log_3 9 = \log_3 3^2 = 2\log_3 3 = 2$.

    Remember log of a number to the same base is equal to 1 and $\log_3 3 = 1$.

6. **Best answer is D**.

    Taking logarithm on both sides and noting that 2 is the common base for 2 and 8,

    $(2x + 8)\log_2 4 = (5x)\log_2 8$;

    $(2x + 8)2 = (5x)3; 4x + 16 = 15x;$

    $11x = 16;$ solving for $x, x = \dfrac{16}{11}$.

7. **Best answer is B**.

    Taking logarithm on both sides and noting that 3 is the common base for 9 and 243,

    $(2x - 8)\log_3 9 = (x)\log_3 243$;

    $(2x - 8)2 = (5x);$

    $4x - 16 = 5x;$

    solving for x, $\quad x = -16$.

8. **Best answer is A**. $\ln e^5 = 5\ln e = 5$.

9. **Best answer is C**. $\ln e^5 - \ln e = 5 - 1 = 4$.

10. **Best answer is C**. By definition, $\dfrac{\log 15}{\log 5} = \log_5 15$.

11. **Best answer is C**. By definition, $\log_4 25 \times \log_3 4 = \log_3 25 = \log_3 5^2 = 2\log_3 5$.

17.   Geometry

At any point in space, there are 360 degrees. That means, when you look at a straight line, there are 180 degrees on either side of a straight line. When two angles add up to 90 degrees, the angles are called complimentary. When two angles add up to 180 degrees, the angles are called supplementary.

SGK's Short Cut:

- A couple of tricks that will help you remember complimentary and supplementary angles.
  - Complimentary = 90 and Supplementary = 180. Remember them in the alphabetical order.
  - Supplementary angle is an angle on a Straight line = 180 degrees. Remember both Supplementary and Straight line start with "S".

Example 1: What is angle x in the adjacent figure if z=40?

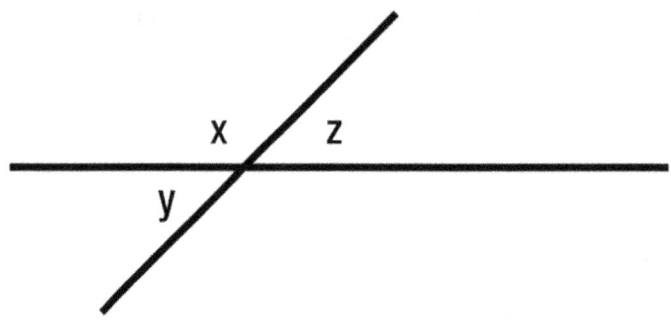

Because x and z are supplementary angles, x+z=180.

x=180-z=180-40 = 140.

Example 2: Continuing from example 1, what is y?

Because x and y are supplementary angles, x+y=180.

y=180-x=180-140=40. Note when two lines intersect, the opposite angles are always equal. Intersecting lines are explained later in this chapter.

Example 3: In the triangle below, what is angle a if b=60?

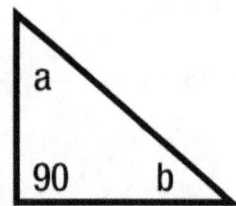

Because a and b are complementary angles, a+b=90.

a=90-b=90-60=30.

Example 4: Continuing from example 3, if a=4b, what is a?

a and b are complimentary. Hence, a+b=90; because a=4b, substitution gives 4b+b=90 or 5b=90. b = $\frac{90}{5}$ = 18. a= 4b = 4×18=72.

Example 5: If x and y are supplementary and are in the ratio of 1:5, what is the larger angle equal to?

Assume x=k, y=5k. The sum of supplementary angles is 180. Therefore, k+5k=180 or 6k=180 and k = 30. Note the problem is asking for the larger angle, which is y = 180-30=150

Example 6: If x and y are supplementary and if x = 4k+1 and y = k+9, what is y?

Sum of supplementary angles is 180. Therefore, 4k+1+k+9=180; 5k+10=180; 5k = 180-10=170. k = $\frac{170}{5}$=34. y=34+9=43.

Intersecting lines

Opposite angles at two intersecting lines are equal. Example 2 illustrates this concept.

Example 7: y and z are opposite vertical angles. If y=4k+1 and z=7k-38, find y.

Opposite angles are equal. Therefore, 4k+1=7k-38. 7k-4k=1+38. 3k = 39 and solving for k, k = $\frac{39}{3}$=13.

y=4k+1=4×13+1=53.

Parallel lines and Transversal

A line intersecting two parallel lines is called a transversal line. It is important to note that there are 8 angles; 4 sets of angles. a, b, g and h are called exterior angles. c, d, e and f are called interior angles.

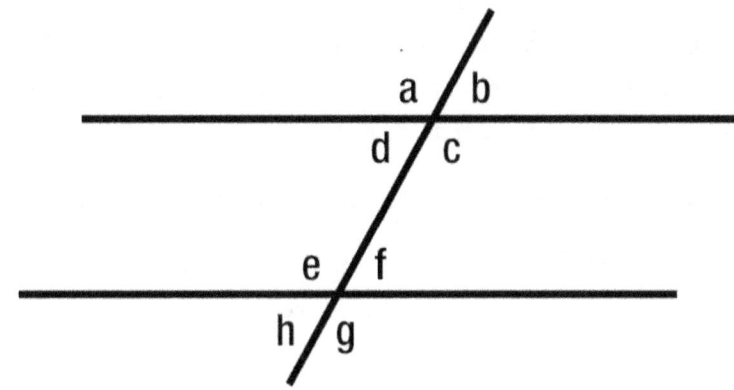

Given opposite angles are equal, a=c, b=d, e=g and f=h.

Given lines are parallel, corresponding angles are equal. a=e, b=f, c=g and d=h.

Which means, a=c=e=g and b=d=f=h.

Opposite interior angles are equal. (c=e and d=f).

Opposite exterior angles are equal. (a=g and b=h).

Example 8: Given a=40 degrees, find the other angles.

a=c=e=g=40 degrees. b= 180-40=140=d=f=h.

Example 9: If a=x+10 and h=3x+50; find b.

b=h (opposite exterior angles). Because a and b are supplementary, they add up to 180.

Hence, x+10+3x+50=180; 4x+60=180; 4x=180-60=120 or x =$\frac{120}{4}$=30.

Hence, a=30+10=40 and b=3x+50=3×30+50=90+50=140.

Triangles

"Tri" means three. A triangle has three sides and three angles. The sum of angles in a triangle is equal to 180.

Example 10: If angles in a triangle are in the ratio of 2:3:5, what is the largest angle?

Let the angles be 2x, 3x and 5x. Sum of the angles = 2x+3x+5x=180; 10x=180 or x=18.

The largest angle is 5x=5×18=90. In addition, you know the largest angle is 90 if sum of two parts equal to the third. In this case, in the ratio 2:3=2+3=5. Hence, the largest angle is 90 degrees.

Properties of a Triangle

- *Opposite to the largest side is the largest angle (and vice versa).*
- *Opposite to the smallest side is the smallest angle (and vice versa).*
- *Each side should be less than the sum of the two other sides.*

Example 11: If two sides of a triangle are 3 and 7, what is the range of values for the third side?

To solve this problem, simply find out the sum and difference of the two sides that are given. i.e., 3+7 and $|7 - 3|$; 10 and 4. Hence, the third side must be between 4 and 10 (not equal to). In other words, the third side should be such that 4 < x < 10.

If x and y are two sides of a triangle, then the third side z should be such that

$$|x - y| < z < x + y.$$

One side, x	Another side, y	$|x - y|$	x+y	Third side, z
2	5	5-2=3	5+2=7	3<z<7
3	7	7-3=4	7+3=10	4<z<10
4	6	2	10	2<z<10
5	5	0	10	0<z<10
1	11	10	12	10<z<12
7	9	2	16	2<z<16

Types of Triangles

A *scalene* triangle is one where all three sides are different. An *isosceles* triangle is one where two sides (and two corresponding angles) are equal. An *equilateral* triangle is one where all three sides (and all three angles) are equal. Each angle in an equilateral triangle is equal to 60 ($\frac{180}{3} = 60$).

Example 12: If ABC is an isosceles triangle where AB=AC and angle A is 40 degrees, what are the other two angles?

Because b=c and b+c = 180-40=140; each of b and c should be equal to $\frac{140}{2} = 70$ degrees

Perimeter of a Triangle

The perimeter of a triangle is the sum of its three sides. Perimeter of an equilateral triangle = 3×a where "a" is the side.

Area of a Triangle

Area of a triangle is found by the formula, area $= \frac{1}{2} \times$ base $\times$ height. Height is defined as the perpendicular drawn from the opposite vertex to the base. In the example below, BC is the base and AD is the height.

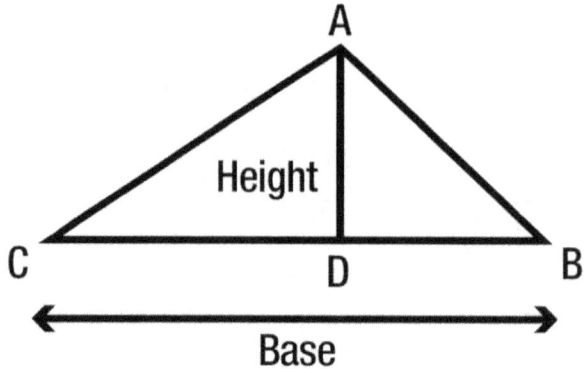

In an equilateral triangle, BD=DC=$\frac{1}{2} \times$ base $= \frac{a}{2}$; where "a" is the side of an equilateral triangle.

Height of an equilateral triangle is $\frac{\sqrt{3}}{2}$a. Hence, area of an equilateral triangle is equal to $\frac{\sqrt{3}}{4}a^2$.

SGK's Short Cut:

- $\frac{\sqrt{3}}{2} = 0.866$. Hence, <u>the height</u> of an equilateral triangle is always <u>less than the side</u>.
- If side has $\sqrt{3}$ in it, height will not have $\sqrt{3}$ in it and vice versa.
- Area of an equilateral triangle always has $\sqrt{3}$.

Using Ratios to Compare Perimeters and Areas of Triangles

Using the example of an equilateral triangle, note that perimeter is proportional to the side whereas area is proportional to the square of the side. What this means is doubling the side of an equilateral triangle will double the perimeter. Doubling the side will increase the area by <u>2 times 2</u>=4 times. Tripling the side will increase the area by <u>3 times 3</u>=9 times. Similarly, if the sides of two equilateral triangles are in the ratio of 2:3, their areas will be in the ratio of 2×2:3×3=4:9.

Similar Triangles

When two triangles are such that the corresponding angles are equal and ratio of corresponding sides are equal, the triangles are said to be "similar triangles". Think of triangles are made to "scale", "scale up" or "scale down". Perimeters of similar triangles are proportional to the sides (equal to the ratio of the corresponding sides). Areas of similar triangles are proportional to the square of the sides (equal to the ratio of squares of the corresponding sides).

<u>Example 13:</u> Triangle ABC and ADE are similar. $\frac{AB}{AD} = 3$. What is $\frac{\text{Area of triangle ABC}}{\text{Area of triangle ADE}}$?

Because $\frac{AB}{AD} = 3$, $\frac{\text{Area of triangle ABC}}{\text{Area of triangle ADE}} = 3 \times 3 = 9$.

Example 14: Continuing from example 13, if $\frac{AB}{AD} = 3$ and DE=4 units, what is BC?

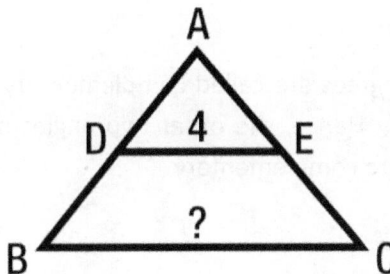

Because $\frac{AB}{AD} = 3 = \frac{BC}{DE}$, BC = 3×DE=3×4=12 units.

Right Triangle

90 degree angle is called a right angle. One of the angles in a right triangle is equal to 90 degrees. Therefore, sum of the other two angles must also be equal to 180-90=90 degrees. A right isosceles triangle has angles 45, 45 and 90 degrees.

The side opposite to the right angle is called the _hypotenuse_. 90 is the largest angle in a right triangle. Hence, _hypotenuse_ is the largest side in a right triangle.

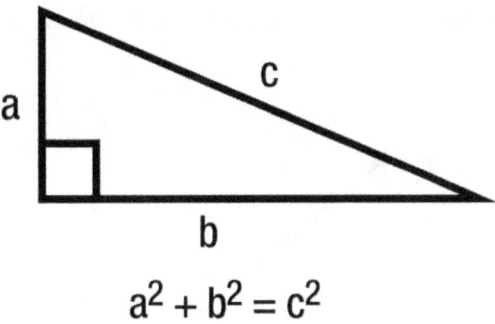

$$a^2 + b^2 = c^2$$

Pythagorean Theorem

In the adjacent figure, "c" is the largest side, hypotenuse of the right triangle, ABC. "a" and "b" are the two smaller sides. As per the Pythagorean Theorem, in a right triangle, square of the hypotenuse is equal to the sum of the squares of the other two sides. i.e., $a^2 + b^2 = c^2$.

SGK's Short Cut: In a right isosceles triangle, two angles and two sides are equal. Hence, a=b which implies that $a^2 + a^2 = c^2; 2a^2 = c^2; a = b = \frac{c}{\sqrt{2}}$ for a right isosceles triangle.

Area of a Right Triangle

Area of a right triangle is $\frac{1}{2}$ product of the two smaller sides = $\frac{1}{2}$ab.

Area of a right isosceles triangle $= \frac{1}{2}a^2 = \frac{1}{4}c^2$.

Complementary Angles

Two angels that add up to 90 degrees are called complementary angles. Note that in a right triangle, one of the angles is 90 degrees. Hence, the other two angles must add up to 180-90 = 90 degrees. Hence, angles in a right triangle are complementary.

Interior Angles and Exterior Angles

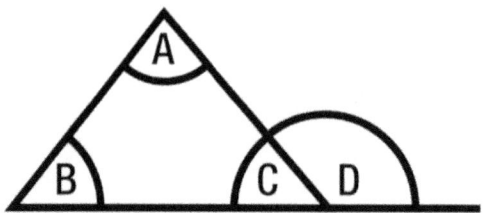

Angles A, B and C are called interior angles. Angle D is called an exterior angle.

Sum of angles A, B and C = 180 degrees (because sum of angles in a triangle = 180 degrees). Sum of angles C and D = 180 degrees (because C and D are supplementary in nature). Hence, we can note that angle D is equal to the sum of the angles A and B. In a triangle, the sum of the interior angles = opposite exterior angle. The sum of all three exterior angles in a triangle equals to 360 degrees.

Trigonometric Ratios

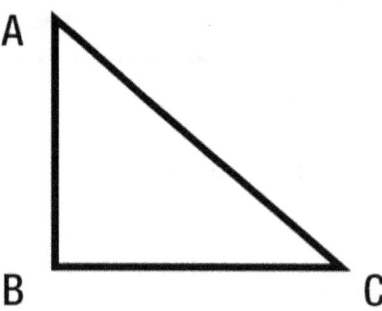

In the right triangle ABC, sin(A) is defined as $\frac{\text{Opposite Side}}{\text{Hypotenuse}} = \frac{BC}{AC}$.

cosine(A) is defined as $\frac{\text{Adjacent Side}}{\text{Hypotenuse}} = \frac{AB}{AC}$.

tangent(A) is defined as $\frac{\text{Opposite Side}}{\text{Adjacent Side}} = \frac{BC}{AB}$.

SGK's Short Cut: Note that hypotenuse is the largest side in a right triangle; Hence, sin(A) and cos(A) can never be greater than 1. On the other hand, tan(A) can be more than 1.

Law of Sines

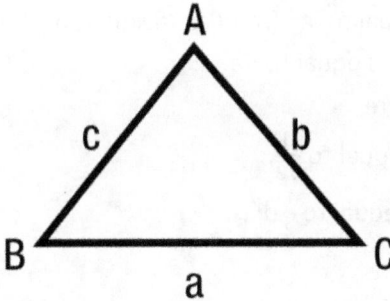

When we are dealing with a general triangle, we have

$$\frac{a}{\sin A} = \frac{b}{\sin B} = \frac{c}{\sin C}.$$

Law of Cosines

$$a^2 = b^2 + c^2 - 2bc \cos A; \ b^2 = a^2 + c^2 - 2ac \cos B; \ c^2 = a^2 + b^2 - 2ab \cos C.$$

When to Use Law of Sines and Cosines

- When two sides and the included angle is given, use Law of Cosines. Otherwise, use the Law of Sines to solve a problem.

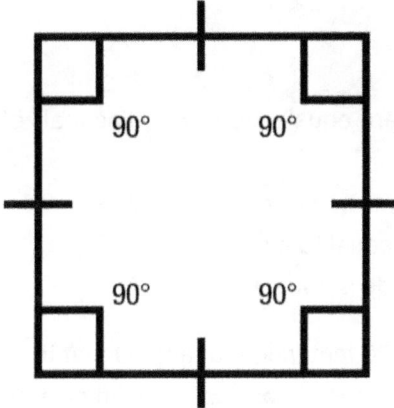

Square

A square has four sides. All four sides and all four angles are equal (AB=BC=CD=AD). The sum of angles in a square = 360 degrees, hence, each angle in a square is equal to 90 degrees.

Properties of a Square

- The diagonals of a square are equal (AC=BD). Diagonals of a square intersect at right angles. The diagonals of a square bisect each other (AE=EC=DE=EB). Hence, diagonals divide a square into 4 equal right isosceles triangles.

- If "a" is equal to the side of a square,
  - Diagonal of a square (AC=BD) is equal to $\sqrt{2}a$.
  - Area of a square is equal to a^2 (read as "a squared", hence the name "square").
  - Perimeter of a square is equal to 4a.
- If "d" is diagonal of a square,
  - Side of the square is equal to $\frac{d}{\sqrt{2}}$.
  - Area of the square is equal to $\frac{1}{2}d^2$.

Rectangle

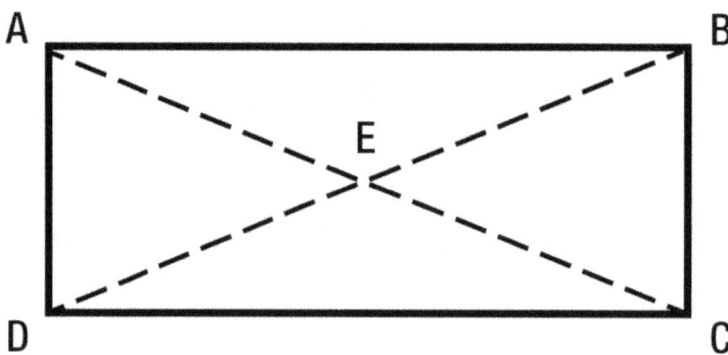

A rectangle is one where opposite sides are equal and perpendicular.

Properties of a Rectangle

- Diagonals of a rectangle are equal. However, diagonals of a rectangle <u>do not</u> intersect at right angles.
- If "a" and "b" are equal to the sides of a rectangle
  - Area of a rectangle is equal to a×b.
  - Perimeter of a rectangle is equal to 2(a+b).

<u>Example 15:</u> A rectangular field is 70 feet in length and 60 feet in width.   A walking trail 6 inch wide is to be built around the rectangular field. If it costs $5 to build each sq.ft, what is the total cost to build the walking trail?

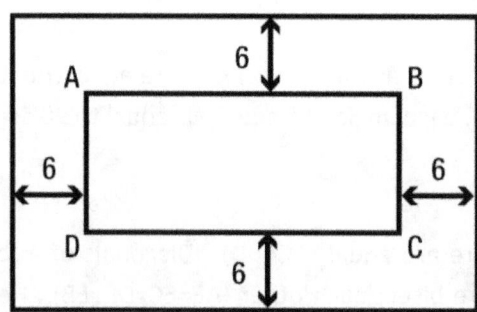

Note that the problem is specifying units in both feet and inches. Make sure you convert inches into feet. 12 inches = 1 feet. Also, note that the fence is on all four sides. Given AB=70 ft. and BC=60 ft., the length of the outer rectangle = 70+0.5+0.5 (to account for both sides) feet = 71 feet. Similarly, the width of the outer rectangle = 60+0.5+0.5 = 61 feet. A common mistake made by students is to add "0.5" instead of adding "0.5+0.5". Area of the walking trail = Area of the outer rectangle − Area of the inner rectangle = $71 \times 61 - 70 \times 60 = 131$ sq. ft. If each sq.ft costs \$5, the cost to build the walking trail = $131 \times \$5 = \$655$.

<u>Example 16:</u> What are the length and width of a rectangle if the area equals to 48 and the perimeter equals to 32?

Let length be l and width be w. It is given that l×w=48 and 2(l+w)=32; or l+w=16. Solving for l and w, l=12 and w=4.

Rhombus

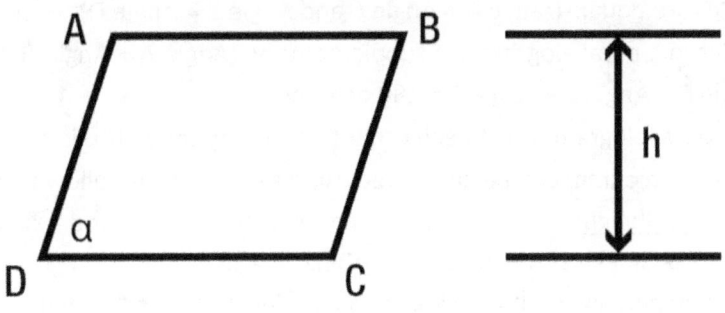

All four sides of a rhombus are equal but not perpendicular (AB=BC=CD=AD).

Properties of a Rhombus

- Opposite sides of a rhombus are parallel (AB is parallel to CD and BC is parallel to AD).
- Opposite angles are equal. (Angle A = Angle C and Angle B = Angle D). Adjacent angles in a rhombus are supplementary. (Angle A + Angle B = Angle C + Angle D = Angle A + Angle D = Angle B + Angle C = 180 degrees).
- The area of a rhombus can be calculated using several of the following methods.
    - <u>Given the diagonals:</u> If the diagonal AC = x and BD = y, the area ABCD = $\frac{1}{2}$xy.
    - <u>Given base and height:</u> If the side of a rhombus is "a" and height (distance between the two parallel lines) is "h", the area of ABCD = a×h.
    - <u>Given the side and angle:</u> If the side of a rhombus is "a" and the angle ADC is α, the area of ABCD = $a^2 \sin \alpha$.

Parallelogram

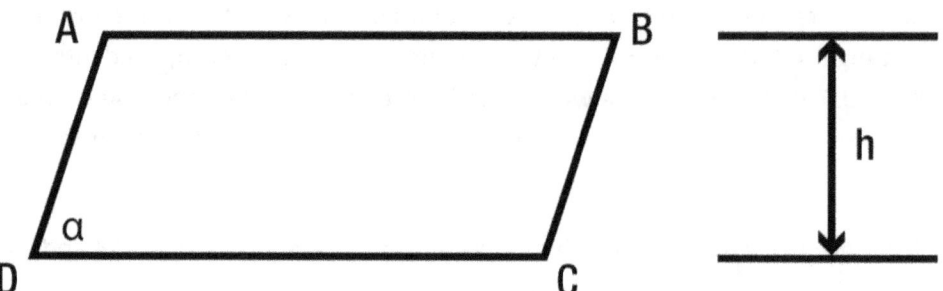

The opposite sides of a parallelogram are parallel and equal but not perpendicular.

Properties of a Parallelogram

- Opposite angles are equal. (Angle A = Angle C and Angle B = Angle D).
- Adjacent angles in a parallelogram are supplementary. (Angle A + Angle B = Angle C + Angle D = Angle A + Angle D = Angle B + Angle C = 180 degrees).
- Diagonals of a parallelogram bisect each other but are not equal. (AC is not equal to BD).
- The area of a parallelogram can be calculated using several of the following methods.
  - <u>Given base and height:</u> If the base of a parallelogram is "b" and height (distance between the two parallel lines) is "h", the area of ABCD = b×h.
  - <u>Given the side and angle:</u> If the sides of a parallelogram are "a" and "b", and the angle ADC is α, the area of ABCD = a×b×sin α.

Trapezoid

Two sides of a Trapezoid are parallel. (AB is parallel to CD).

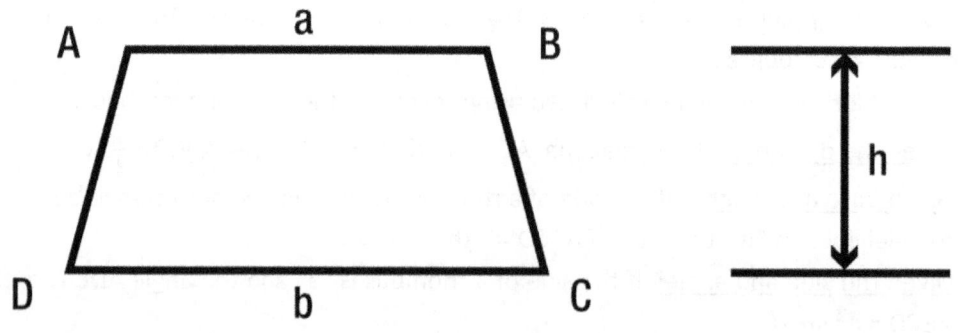

The area of a trapezoid is equal to $\frac{1}{2}(a + b)h$ where a and b are the parallel sides and h is the distance between the parallel sides.

Geometry Problems Involving Ratios

Quite often you run into problems as illustrated below.

Example 17: The side of a square is doubled. What is the ratio of the area of the new square to the previous one?

Assume side of the square is 3 units (or some other number you like). Area is 3×3 = 9 units. If the side is doubled, the new side will be 3×2 = 6 units. Area of the new square is 6×6 =36 units. The ratio of area of the new square to the previous one is 36:9 = 4:1.

SGK's Short Cut: In general, the area of a square is proportional to the square of the side. So, if the side is doubled, area increases by 2×2 = 4 times. If the side is tripled, the area of the square increases by 3×3=9 units. If the side is increased by 5 times, the area increases by 5×5=25 times.

Example 18: If the length of a rectangle is doubled and the width is tripled, what is the new area of the rectangle?

X = the length of the original rectangle and Y = the width of the original rectangle.

The area of the original rectangle = XY.

The length of the new rectangle = 2X and the width of the new rectangle = 3Y

New area = 2X×3Y = 6XY = 6 times the area of the original rectangle.

Hexagon

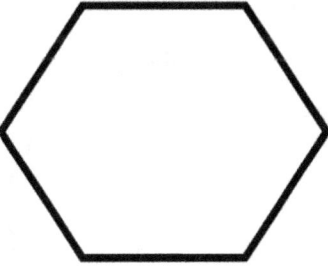

A hexagon is a polygon with 6 sides. A regular hexagon is one where all sides are equal and all angles are equal. A regular hexagon can be divided into 6 equilateral triangles. If "a" is the side of a hexagon, the area of the hexagon is equal to $6 \times \frac{\sqrt{3}}{4} \times a \times a = \frac{3\sqrt{3}}{2}a^2$. Note also that the area of a hexagon has $\sqrt{3}$ (just like area of an equilateral triangle has).

Angles in a Polygon

The sum of the interior angles in a polygon = $(n-2) \times 180$ where is "n" is the number of sides.

For a regular polygon (one where all sides and angles are equal), interior angle is calculated as $(n-2)\frac{180}{n}$.

- For a Triangle, n = 3. The sum of interior angles = (3-2) × 180 = 180 degrees.
- For a quadrilateral, n = 4. The sum of interior angles = (4-2) × 180 = 360 degrees.
- For a pentagon, n=5. The sum of interior angles = (5-2)×180 = 3.180 = 540 degrees.
  - For a regular pentagon, each interior angle is equal to $\frac{540}{5} = 108$ degrees.
- For a hexagon, n=6. The sum of interior angles = (6-2)×180 = 4×180 = 720 degrees.
  - For a regular hexagon, each interior angle is equal to $\frac{720}{6} = 120$ degrees.

The exterior angle of a regular polygon = $\frac{360}{n}$, where "n" is the number of sides.

How are the interior and exterior angles in a regular polygon related? The interior and exterior angles in a regular polygon are supplementary. Hence, the sum of an interior and exterior angles of a regular polygon is equal to 180 degrees.

Example 19: Find the interior and exterior angles of polygon with 12 sides. (n=12).

The exterior angle = $\frac{360}{n} = \frac{360}{12} = 30$ degrees.

The interior angle = 180-30 = 150 degrees.

Note that as "n" increases, the exterior angle gets smaller and smaller. Hence, the interior angle increases as n increases.

Circles

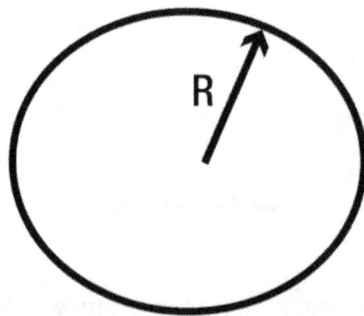

Each point on a circle is equidistant from the "center" of the circle. The distance between any point on the circle and the center is called the "radius" of the circle. A line that passes through the center of the circle is called the "diameter" of the circle. Diameter = 2×radius. In a circle, there are 360 degrees (very important to note this).

Circumference of a circle = $2\pi r = \pi d$, where r is the radius and d is the diameter.

Area of a circle = $\pi r^2 = \frac{\pi}{4} d^2$, where r is the radius and d is the diameter.

Sector

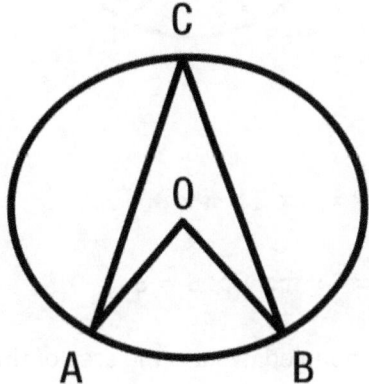

A sector is a portion of a circle. A sector includes an angle called central angle (Angle AOB). Length and area of a sector are proportional to the central angle.

A sector with 60 degrees will have $\frac{60}{360} = \frac{1}{6}$ th the circumference and $\frac{1}{6}$ th the area of a full circle.

A sector with 40 degrees will have $\frac{40}{360} = \frac{1}{9}$ th the circumference and $\frac{1}{9}$ th the area of a full circle.

Inscribing angle is half the central angle (Angle ACB). Angle AOB = 2 × Angle ACB.

Because a semi-circle (half circle) has 180 degrees, inscribed angle in a semi-circle is always 90 degrees.

<u>Example 20:</u> If AB is diameter of the circle, what is angle ABC?

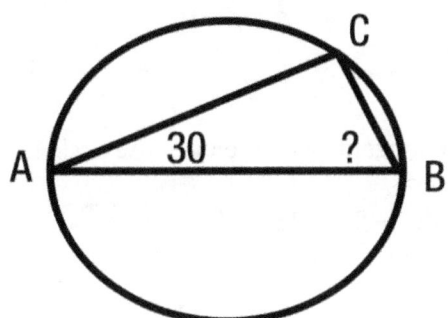

Because Angle C is inscribed angle in a semi-circle, angle C = 90 degrees. Angle B = 90-30 = 60 degrees.

Example 21: Find the length and area of a sector as shown in the diagram where AB is 8 units.

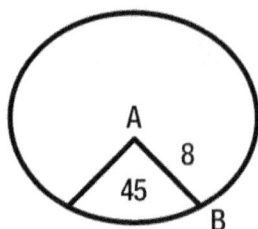

The ratio is $\frac{45}{360} = \frac{1}{8}$.

Length of the sector = $\frac{45}{360} \times 2 \times \pi \times 8 = \frac{1}{8} \times 2 \times \pi \times 8 = 2\pi$.

Area of the sector = $\frac{45}{360} \times \pi \times 8 \times 8 = \frac{1}{8} \times \pi \times 8 \times 8 = 8\pi$.

Example 22: If the radius of a circle is doubled, what is the area of the new circle?

Area of a circle is proportional to the square of the radius. If the radius is doubled, the area increases by 2×2 = 4 times.

Tangent

A tangent line is a straight line that touches the circle in one and only one point. The radius connecting the center of the circle and the point of tangency is always perpendicular to the tangent line.

Radius of a circle, length of the tangent and the distance between the center of a circle and the external point are related by the Pythagorean Theorem.

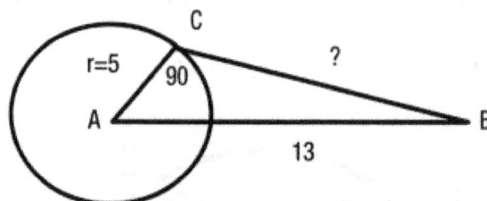

Example 23: Find the length of the tangent given radius of the circle is 5 units and AB = 13 units.

Using the Pythagorean Theorem, $BC^2 + 5^2 = 13^2$; $BC = \sqrt{169 - 25} = \sqrt{144} = 12$.

Pair of Circles and Tangency

Concentric circles		Number of tangents = 0
Intersecting Circles One point of intersection (Internal)		Number of tangents possible = 1
Intersecting Circles Two points of intersection		Number of tangents possible = 2
Intersecting Circles One point of intersection (External)		Number of tangents possible = 3
Non-Intersecting Circles		Number of tangents possible = 4

Polygons within a Circle

Study the following polygons within a circle.

An Equilateral Triangle within a Circle

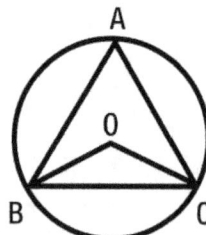

Note that the central angle BOC is equal to $\frac{360}{3} = 120$ degrees.

If radius of the circle is "r" units, the length of the equilateral triangle = 2r cos 30 = $2r\frac{\sqrt{3}}{2}$ = $\sqrt{3}$r.

A Square in a Circle

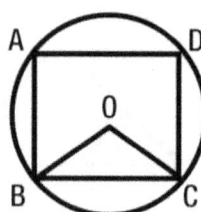

Central angle BOC is equal to $\frac{360}{4} = 90$ degrees. Hence, each of the angles in the triangle OBC=OCB=45 degrees. If "r" is the radius of the circle, the length of side of the square = $\sqrt{2}$ r units.

A Hexagon in a Circle

Central angle is equal to $\frac{360}{6} = 60$ degrees. Each triangle is an equilateral triangle. Hence, each side of the triangle = each side of the hexagon = radius of the circle = r.

Properties of a Cube

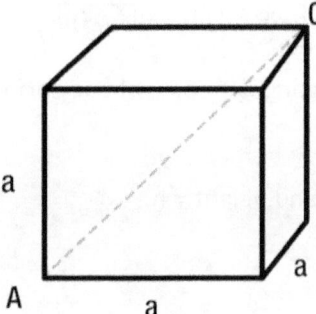

Each side in a cube is equal. All edges in a cube intersect at an angle of 90 degrees. Cube has 6 faces, 8 vertices and 12 edges. Each face has an area of a^2 where "a" is the side of the cube. The volume of the cube is a^3 units. The surface area of a cube is equal to $6a^2$.

The longest diagonal in a cube is the one that connects the opposite vertices (points A and C in the diagram). It is calculated as

$$\sqrt{a^2 + a^2 + a^2} = \sqrt{3a^2} = \sqrt{3}a.$$

Properties of a Rectangular Solid

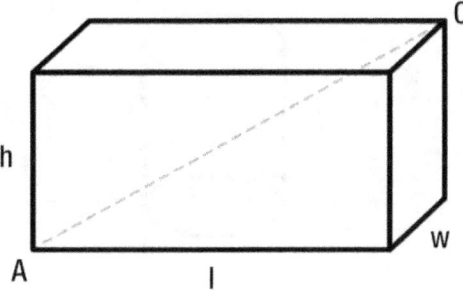

A rectangular solid is one where all three sides are not equal (two sides can be equal). A rectangular solid has 6 faces, 8 vertices and 12 edges. Opposite faces have equal area. Hence, the surface area of a rectangular solid is equal to $2(lw + lh + wh)$, where l = length, w = width and h = height.   Volume of the rectangular solid is equal to <u>lwh</u>. The longest diagonal in a rectangular solid is $\sqrt{l^2 + w^2 + h^2}$.

Calculate Volume of a Rectangular Solid Given the Surface Areas

One can calculate the volume of a rectangular solid, given the surface areas. Let us say, the three sides of a rectangular solid are length = l, width = w and height = h.

Let us say, the surface areas lw, wh and lh are given.

Multiply all three surface areas to get = lw×wh×lh = $l^2w^2h^2$.

To calculate the volume, take the square root, lwh = $\sqrt{l^2 w^2 h^2}$.

Example 24: Calculate volume of a rectangular solid whose surface areas are 24 square units, 8 square units and 12 square units.

Say, the sides are length = l, width = w and height = h.

l×w = 24, w×h =8 and l×h = 12.

Volume = lwh = $\sqrt{24 \times 8 \times 12}$ = 48 cubic units.

In fact, the sides can now be determined as follows.

Length, l $= \dfrac{lwh}{wh} = \dfrac{48}{8} = 6$ units.

Width, w $= \dfrac{lwh}{lh} = \dfrac{48}{12} = 4$ units .

Height, h $= \dfrac{lwh}{lw} = \dfrac{48}{24} = 2$ units.

Cylinder

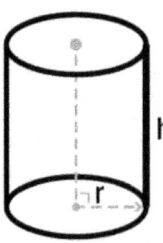

A cylinder is defined by height, h and radius, r. The surface area of a cylinder is $2\pi r^2 + 2\pi rh$ when we count both top and bottom of the cylinder. Sometimes, the problem might ask you to calculate the surface area without the top of the cylinder. The surface area without the top and bottom of the cylinder is $2\pi rh$. The volume of the cylinder is $\pi r^2 h$.

Sphere

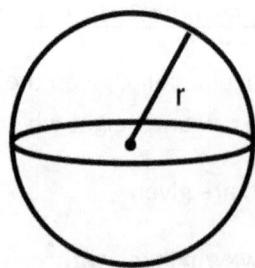

A sphere is defined by radius, r.

The volume of a sphere = $\frac{4}{3}\pi r^3$.

The surface area of a sphere = $4\pi r^2$.

The ratio of volume to the surface area of a sphere = $\frac{1}{3}r$.

Coordinate Geometry

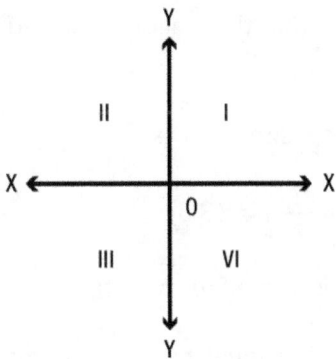

To answer problems involving coordinate geometry, note the following:

- x-axis and y-axis act as references. They intersect each other at right angles.
- The point of intersection of x and y axis is called the origin. The origin has the coordinates of (0,0).
- Right of x-axis holds positive values and left x-axis holds negative values.
- Top of y-axis holds positive values and bottom of y-axis holds negative values.
- Any point in the plane is represented by coordinate pair (x,y); where x represents distance along x-axis (or distance from y-axis) and y represents distance along y-axis (or distance from x-axis).
- Points on the x-axis have coordinates of (x,0) and Points on the y-axis have coordinates of (0,y).
- A plane is divided into four quadrants. In the first quadrant, both x and y coordinates are positive. In the second quadrant, x coordinate is negative and y coordinate is positive. In the third quadrant, both x and y coordinates are negative. In the fourth quadrant, x coordinate is positive and y coordinate is negative.

Mid-Point Formula

The mid-point of two points (x_1, y_1) and (x_2, y_2) is equal to $\left(\frac{x_1+x_2}{2}, \frac{y_1+y_2}{2}\right)$. The mid-point is equvidistant from the two points. Calculation of the mid-point involves the average formula. The coordinates of the mid-point are in between the coordinates of the two points.

Example 25: If (1,3) and (7,17) are two points along the diameter of a circle, what are the coordinates of the center of the circle?

Note the center of the circle is the mid-point on the diameter; the coordinates of the center of the circle are $\left(\frac{1+7}{2}, \frac{3+17}{2}\right) = (4,10)$. Note that x-coordinate 4 is equvidistant from 1 and 7. Similarly, y-coordinate 10 is equvidistant from 3 and 17.

Example 26: If three vertices of a square have the coordinates (1,1), (1,5) and (5,1), what are the coordinates of the fourth vertex?

Method 1: The concept to apply here is that the diagonals of a square bisect (divide in two equal parts) each other.

Step 1: Identify (1,5) and (5,1) as the opposite vertices.

Step 2: Calculate the midpoint as $\left(\frac{1+5}{2}, \frac{5+1}{2}\right) = (3,3)$.

Step 3: Assume (x,y) are the coordinates of the fourth vertex.

Step 4: The mid-point of (x,y) and (1,1) should be same as the mid-point of (1,5) and (5,1) due to symmetry. Hence, $\left(\frac{x+1}{2}, \frac{y+1}{2}\right) = (3,3)$. Solving for x and y, x=5 and y=5. The fourth vertex is (5,5).

SGK's Short Cut: A short cut to this problem is to realize that the sum of x-coordinates and y-coordinates of opposite vertices are equal. That means, 1+X=1+5 and 1+Y=5+1 or X=5 and Y=5. See below.

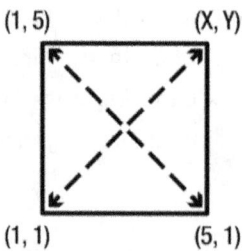

(1, 5)   (X, Y)

(1, 1)   (5, 1)

Distance Formula

The distance between two points (x_1, y_1) and (x_2, y_2) is equal to $\sqrt{(x_2 - x_1)^2 + (y_2 - y_1)^2}$.

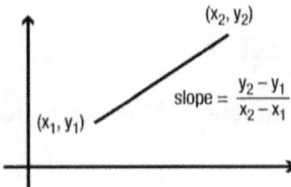

Example 27: What is the distance between the points (1,3) and (5,9)?

Use the formula, $\sqrt{(x_2 - x_1)^2 + (y_2 - y_1)^2} = \sqrt{(5-1)^2 + (9-3)^2} = \sqrt{4^2 + 6^2} = \sqrt{16 + 36} = \sqrt{52}$ units.

Slope of a Line

Slope of a line is related to the angle made by the line with x-axis.

The slope of a line joining two points (x_1, y_1) and (x_2, y_2) is equal to $\frac{y_2 - y_1}{x_2 - x_1}$.

- Slope also indicates the rate of change y with respect to the rate of change in x.
- Higher the slope, higher the rate of change in y with respect to unit change in x.
- A higher slope also means the line is more slanted (rises up more quickly).
- The slope of x-axis is zero and the slope of y-axis is not determined.
- Parallel lines have the same slope.
- Product of slopes equals to -1 when the lines are perpendicular.

Example 28: What is the slope of the line that passes through the points (1,4) and (2,7)? What is the slope the line that is perpendicular to this line?

The slope of the line $= \frac{y_2 - y_1}{x_2 - x_1} = \frac{7-4}{2-1} = 3$. The slope of the perpendicular line is negative reciprocal or $-\frac{1}{3}$.

Equation of a Straight Line

Slope-intercept form: y=mx+c; where m is the slope and c is the y-intercept.

General form: ax+by+c=0. The slope of this line is $\frac{-a}{b}$ and y-intercept is $\frac{-c}{b}$.

The equation of x-axis is y=0. The equation of a line parallel to x-axis is y=c.

The equation of y-axis is x=0. The equation of a line parallel to y-axis is x=c.

Calculate the Slope Given an Equation

Rewrite the equation in slope-intercept form if it is not already in the form. Note m as the slope. OR rewrite the equation in general form and note $\frac{-a}{b}$ as the slope.

Example 29: What is the slope of the line represented by the equation 2x+3y+5=0?

The slope of the line is $\frac{-a}{b} = \frac{-2}{3}$.

Equation of a Line Passing Through Origin

y=mx is the equation of line passing through the origin.

Given the equation of a Line, Find x-Intercept

To find x-intercept, substitute y=0 in the equation. Then, solve for x.

Given the equation of a Line, Find y-Intercept

To find y-intercept, substitute x=0 in the equation. Then, solve for y.

Example 30: Find the x and y intercepts of the line 2X+3Y=12.

X-Intercept: When Y=0, 2X=12 and X=$\frac{12}{2}$ = 6. Hence, X=6 is the x-intercept.

Y-Intercept: When X=0, 3Y=12 and Y=$\frac{12}{3}$ = 4. Hence, Y=4 is the y-intercept.

Finding Area of a Triangle with Origin as One of the Vertices

Area of a triangle with vertices (0, 0), (x_1, y_1) and (x_2, y_2) is $\frac{1}{2}|x_1y_2 - x_2y_1|$.

Example 31: What is the area of the triangle formed by vertices (0,0), (1,4) and (9,8)?

Area of the triangle is $\frac{1}{2}|x_1y_2 - x_2y_1| = \frac{1}{2}|1 \times 8 - 4 \times 9| = \frac{1}{2}|8 - 36| = \frac{1}{2}|-28| = 14$ units.

Finding Area of a Triangle Given Three Coordinates

First convert one of the vertices to origin. Then, apply the above formula.

Example 32: What is the area of the triangle formed by vertices (1,1), (2,5) and (7,3)?

First we have to convert one of the vertices as origin. Let us say, we convert (1,1) as origin. That means, we need to subtract (1,1) from all three vertices. New vertices are (0,0), (1,4) and (6,2). Now applying the formula, area of the triangle is $\frac{1}{2}|x_1y_2 - x_2y_1| = \frac{1}{2}|1 \times 2 - 4 \times 6| = \frac{1}{2}|2 - 24| = \frac{1}{2}|-22| =$ 11 units.

Finding Area of a Square Given Opposite Vertices

Find the length of the diagonal using the distance formula. Then, use the formula, area of the square = $\frac{1}{2}d^2$, where d is the diagonal.

Example 33: What is the area of a square with opposite vertices as (1, 4) and (4, 8)?

The length of the diagonal = $\sqrt{(4-1)^2 + (8-4)^2} = \sqrt{3^2 + 4^2} = \sqrt{9 + 16} = \sqrt{25} = 5$ units.

The area of the square = $\frac{1}{2} \times 25 = 12.5$ units.

Practice Problems

1. If triangle ABC is an equilateral triangle, what is angle x?
   A. 30   B. 60   C. 90   D. 120   E. 150

2. In triangle ABC, AB=AC. Angle A =30. What is angle x?
   A. 30   B. 60   C. 90   D. 75   E. 105

3. An equilateral triangle is made with match sticks of length 4 units. What is the area of the triangle?
   A. 12   B. $4\sqrt{3}$ C. $\sqrt{3}$ D. 16   E. 20

4. My son made an equilateral triangle with match sticks of length 5 units. What is the perimeter of the triangle?
   A. 15   B. $25\sqrt{3}$ C. $5\sqrt{3}$ D. 20   E. 12.5

5. What is the side of an equilateral triangle with area equal to $16\sqrt{3}$.
   A. 4   B. 8   C. 16   D. $4\sqrt{3}$ E. $2\sqrt{3}$

6. My daughter drew a square with a diagonal equal to 10 units. What is the area of the square?
   A. 100   B. 200   C. 400   D. 50   E. 150

7. A cubical box is made with each side equal to 6 units. What is the length of the longest diagonal?
   A. $6\sqrt{3}$ B. 12   C. 18   D. $6\sqrt{2}$ E. $12\sqrt{2}$

8. A rectangular box is made with sides 2, 4 and 6 units. What is the length of the longest diagonal?
   A. $6\sqrt{3}$ B. 12   C. $2\sqrt{14}$ D. $6\sqrt{2}$ E. $12\sqrt{2}$

9. When a side of square is tripled, what is the ratio of the area of the new square to the area of the old square?
   A. $\frac{1}{3}$ B. 3   C. $\frac{1}{9}$ D. 9   E. No change

10. My son draws a circle with radius "r" units. My daughter draws another circle with twice the radius. What is the ratio of area of the two circles?
    A. $\frac{1}{9}$ B. 4   C. $\frac{1}{2}$ D. 2   E. π

11. An equilateral triangle is made by using one match stick as a side. By how much is the area of the triangle increased if we use two match sticks instead of one to form each side of the triangle?
    A. $\frac{1}{2}$ B. 2   C. $\frac{1}{4}$ D. 4   E. No change

12. If two sides of a triangle are 4 and 9 units, which of the following cannot be a side of the triangle
    A. 6   B. 7   C. 8   D. 5   E. 11

13. In a right triangle, hypotenuse is 6 units and one of the sides is 3 units. What is the area of the triangle?

A. $\frac{9}{2}\sqrt{3}$ B. 9   C. 18   D. $3\sqrt{27}$ E. 27

14. In a triangle, hypotenuse is 13 and one of the sides is 12. What is the perimeter of the triangle?
    A. 30   B. 60   C. 25   D. 65   E. 15

15. If a side of a square is increased by 20%, area is increased by what percent?
    A. 30%   B. 20%   C. 40%   D. 44%   E. 50%

16. What is interior angle of a regular octagon? (n=8)
    A. 45   B. 120   C. 135   D. 150   E. 30

17. What is exterior angle of a regular polygon with 18 sides?
    A. 20   B. 30   C. 160   D. 150   E. 90

18. If the length of a rectangle increases 10% and width increases by 20%, approximately what
    percentage will the area of the rectangle increase?
    A. 30%   B. 33%   C. 32%   D. 132%   E. 20%

19. What is the volume of a cube whose surface area is 18?
    A. 9   B. 27   C. $3\sqrt{3}$ D. 81   E. $9\sqrt{3}$

20. What is the area of the triangle formed by the vertices (2,3), (5,9) and (-3,-11)?
    A. 9   B. 27   C. $3\sqrt{3}$ D. 6   E. $9\sqrt{3}$

21. ABCD is a rectangle with vertices A=(7,2), B=(9,5) and C=(3,9). Find the coordinates of D.
    A. (9,14) B. (6,1)   C. (1,6)   D. (6,6)   E. (1,1)

22. EFGH is a rectangle with vertices E=(10,2), F=(12,5) and H=(7,4). Find the coordinates of G.
    A. (9,7)   B. (7,1)   C. (7,9)   D. (1,2)   E. (1,1)

23. What is the slope of the line represented by the equation 9x+5y=4?
    A. $\frac{4}{9}$ B. $\frac{4}{5}$ C. $\frac{5}{9}$ D. $\frac{9}{5}$ E. $\frac{-9}{5}$

24. What is the slope of the line ax+3y=2a if it passes through the point (1,3)?
    A. 9   B. -9   C. 3   D. -3   E. $\frac{-1}{3}$

25. If AB is a diagonal of a circle with center (2,3) and if the coordinates of A are (1,8), what are the
    coordinates of B?
    A. (0,9)   B. (1,-4)   C. (3,-2)   D. (2,-3)   E. (1.5,5.5)

26. What is the x-intercept of the line 2x-5y=10?
    A. (2,0)   B. (0,5)   C. (5,0)   D. (5,2)   E. (2,5)

27. What is the area of the triangle formed by the origin and x and y intercepts of the line 2x-5y=10?
    A. 100   B. 10   C. 20   D. 5   E. 40

134

Solutions to Practice Problems

1. **Best answer is D.** ABC is equilateral triangle. Each angle is 60 degrees. x+60=180; x=180-60-120.

2. **Best answer is E.** Because AB=AC; angle B=angle C. B+C=180-30=150. Hence, B and C should be equal to 75 degrees. x=180-75=105.

3. **Best answer is B.** Eliminate choices A, D and E because they do not have radical 3. Area of an equilateral triangle is $\frac{\sqrt{3}}{4}a^2$, where a is the side. Substituting a=4, area = $4\sqrt{3}$.

4. **Best answer is A.** It is a trick question. Perimeter=sum of the sides=5+5+5=15.

5. **Best answer is B.** $\frac{\sqrt{3}}{4}a^2 = 16\sqrt{3}$; solving for a, $a = 8$.

6. **Best answer is D.** Note that all choices can be eliminated except choice D because the side of the square will be less than 10 units (because the diagonal is 10) and the area will be less than 100 units. Otherwise, just use the formula, area = $\frac{1}{2}d^2$, where d is the diagonal. $\frac{1}{2} \times 10 \times 10 = 50$ units.

7. **Best answer is A.** Use the formula $\sqrt{3}a$. $a = 6$, hence the answer is $6\sqrt{3}$.

8. **Best answer is C.** The longest diagonal = $\sqrt{2^2 + 4^2 + 6^2} = \sqrt{4 + 16 + 36} = \sqrt{56} = 2\sqrt{14}$.

9. **Best answer is D.** Let us say the side is 2. When it is tripled, the new side is 6. Ratio of the areas = $\frac{6\times6}{2\times2} = 3 \times 3 = 9$.

10. **Best answer is B.** When radius is doubled, area will increase by $2 \times 2 = 4$.

11. **Best answer is D.** When the side is doubled, area will increase by 2×2=4 times.

12. **Best answer is D.** If two sides are 4 and 9, the third side should be between (9-4) and (9+4). Hence, the third should be greater than 5 and less than 13 units. D is the correct choice.

13. **Best answer is A.** The other leg is calculated using the Pythagorean Theorem. $3^2 + x^2 = 6^2; x = \sqrt{36 - 9} = \sqrt{27} = 3\sqrt{3}$. Area = $\frac{1}{2}3 \times 3\sqrt{3} = \frac{9}{2}\sqrt{3}$.

14. **Best answer is A.** The other leg is calculated using the Pythagorean Theorem. $12^2 + x^2 = 13^2; x = \sqrt{25} = 5$. Perimeter = 5+12+13=30 units.

15. **Best answer is D.** Let the side by 10 units. New side = 10+2=12 units. New Area=12.12=144 units. Increase in area = 144-100=44 units. Hence, new area is 44% more than the previous area.

16. **Best answer is C.** Exterior angle is $\frac{360}{n}$. Exterior angle for an octagon = $\frac{360}{8}$=45 degrees. Hence, the interior angle = 180-45=135 degrees.

17. **Best answer is A.** Exterior angle is $\frac{360}{n}$. Exterior angle for an octagon = $\frac{360}{18}$=20 degrees.

18. **Best answer is C.** New area will be 1.1×1.2=1.32 or 32% more than the previous one.

19. **Best answer is C.** Surface area is $6a^2 = 18$; solving for a, $a = \sqrt{3}$. Volume = $a^3 = 3\sqrt{3}$.

20. ***Best answer is D***. _SGK's Short Cut:_ <u>Step 1:</u> Transform vertex (2,3) to (0,0). Vertex (5,9) becomes (5-2,9-3)=(3,6). Vertex (-3,-11) becomes (-3-2,-11-3)=(-5,-14). New triangle is (0,0), (3,6) and (-5,-14). <u>Step 2:</u> The area of the triangle is $\frac{1}{2}|(3\times-14)-(6\times-5)|=\frac{1}{2}|-42+30|=$ $\frac{1}{2}\times12=6$ units.

21. ***Best answer is C***. <u>Method 1:</u> Mid-point of AC is $\left(\frac{7+3}{2},\frac{2+9}{2}\right)=(5,5.5)$. Let (x,y) are coordinates of D. Mid-point of BD is $\left(\frac{9+x}{2},\frac{5+y}{2}\right)$. Because AC and BD (diagonals) bisect each other, they should have the same mid-point. Hence, $\left(\frac{9+x}{2},\frac{5+y}{2}\right)=(5,5.5)$. Solveing for x and y, $x=1$ and $y=6$.

SGK's Short Cut: Even easier. The sum of x and y coordinates of opposite vertices must be equal. Hence, x+9=7+3; and y+5=2+9. Hence, x=10-9=1 and y=11-5=6.

22. ***Best answer is A***. Let the coordinates of G be (x,y). x+10=12+7 and y+2=5+4. Hence, x=9 and y=7.

23. ***Best answer is E***. Rewriting the equation in slope-intercept form, $y=\frac{9}{5}x+\frac{4}{5}$. Slope $=\frac{-9}{5}$.

24. ***Best answer is D***. (1,3) must satisfy the equation. Substitute (1,3) into the equation. a+9=2a. Solving for a, a=9. The equation is 9x+3y=18. Slope of the line $=\frac{-9}{3}=-3$.

25. ***Best answer is C***. Let the coordinates of B are (x,y). $\left(\frac{1+x}{2},\frac{8+y}{2}\right)=(2,3)$. $1+x=4$ and $8+y=6$. $x=3$ and $y=-2$.

26. ***Best answer is C***. y=0 on x-intercept. Substituting y=0 in the equation, 2x=10 and x=5. X-intercept is (5,0).

27. ***Best answer is D***. X-intercept is (5,0). Y-intercept is $\frac{10}{-5}$=-2. i.e., (0,-2). We need to calculate the area of the triangle (0,0), (5,0) and (0,-2). Area of the triangle $=\frac{1}{2}\times5\times2=5$ units.